A Bigger Problem than Climate Change

The End of Oil

Vernon Coleman

A Bigger Problem than Climate Change

The End of Oil

Dedication

When the lights goes out and the world goes cold you must reach out for the one you love most. In the dark and the chill of the coming long night I'll be holding hands with Donna Antoinette.

Foreword to the Digital Edition

I recently re-read a book of mine which I published in 2007 under the title `Oil Apocalypse'. I was surprised and delighted to see that the book was still extraordinarily relevant.

The book explains the history of peak oil and details events which are behind current energy policies.

To my delight, `Oil Apocalypse' was extremely well received by oil company insiders, many of whom felt it offered an accurate picture of the past, the present and future. A number of executives with great experience in the oil industry sent very complimentary letters.

The book sold very well for a year or so but rather sadly (for me at any rate) it was trashed online by friends of a Green campaigner who was, coincidentally, trying to sell his own book on peak oil. The campaigner wrote a piece for the internet in which he somehow managed both to attack the book and to boast that he hadn't read it and didn't intend to it. And that pretty well finished the book. (The principle of attacking a book you haven't even read seemed to me a novel one then, but today it is relatively common.)

However, the facts in that book are enormously relevant today. They help to explain, for example, why governments are putting so much effort into controlling the use of oil and gas, why subsidies are being provided (not always wisely) for alternative energy programme and why, for example, the European Union has demonised plastics unnecessarily and inaccurately. (The real reason has nothing to do with the environment but is a simple consequence of the fact that we are running out of oil).

It seemed to me that the book was worth restoring and so here it is.

I haven't altered a word in the book which I have given a more appropriate and relevant title.

My views on what has happened more recently in the world of

energy (and the way in which misinformed Greens have forced governments and agencies to adopt inappropriate policies) are contained in my Diaries (notably `Tickety Tonk') and it would seem inappropriate to include them here as well.

In brief, I do not believe that the world will be destroyed by climate change. I do, however, believe that the world as we know it will be destroyed by the oil running out.

The Greens (and the governments they control) have created unnecessary poverty and hunger, have brought forward the day when the problems start to affect us and have done nothing to prepare people for the future which awaits coming generations.

Vernon Coleman, April 2019

`Our daily news sources, newspapers and television, are now so craven, so unvigilant on behalf of the American people, so uninformative, that only in books do we learn what's really going on.'
Kurt Vonnegut (`A Man Without A Country', 2006)

Preface

The oil is running out and, as a result, our civilisation is reaching its end.

The immediate future is not a rosy one. We will face our worst nightmares not when the oil runs out completely but as it runs out.

And the oil will run out slowly but steadily over a decade or more.

But there is hope.

There is always hope.

Maybe we can create a new world, a more restful, peaceful world salvaged out of the debris of the industrial age.

Maybe we can create a simpler, ecologically sound society.

Maybe we can create a world in which there will be less stress and more fulfilment.

This outcome will depend upon our leaders showing strength, wisdom and courage.

In the pessimistic (but, perhaps, practical) belief that our leaders will fail us I have included a blueprint to help you create a survival plan for yourself and your family.

The message in this book is undeniably a gloomy one. The truth can sometimes be alarming. And I doubt if you will ever read a book more alarming than this one. The disaster inexorably heading our way will make any natural disaster, any tsunami, seem trivial in comparison.

And so I felt I should leaven the alarm, and alleviate some of the inevitable despondency, by explaining what practical steps you can take to help give yourself, and your family, the best possible chance of survival.

This book is so utterly terrifying that as I wrote it I had to come to terms with a new vision of the world.

Researching and writing this book changed the way I see the world and made me realise that every political discussion we hear or see today is totally irrelevant. This book is based on facts but the interpretations are, inevitably, my own.

I believe that our civilisation is reaching its end.

And after that there really isn't any more to say.

If you want to know the truth, and you think you can deal with it, turn the page and read this book.

If you don't want to know the truth close the book now and pick up something else.

Vernon Coleman July 2007

Chapter One: Our Unhealthy Addiction To A Gift Of Nature

Your life is about to change beyond recognition. By the time you have finished reading this book your view of the future will have changed irrevocably. Fasten your seat belt. It's going to be a scary read.

If I told you that your house was going to burn down at some time in the next few years would you want to know more? Would you want to make sure that your insurance was up to date? What you want to know how to prepare for the coming disaster?

That's what I thought.

And it's why I've written this book.

The biggest threat to our future isn't from terrorism or global warming.

In comparison to the coming problem I'm going to describe in this book those are trivial and almost inconsequential threats.

I have never been an end-of-the-world-is-coming sort of writer. I don't make a habit of making awful predictions. I've never been in the business of scaring my readers. On the contrary, when the Government and the entire medical establishment were warning that AIDS was going to kill us all I was the lone voice trying to calm everyone down and put things in perspective. (I was widely vilified for that - as I always have been when my medical, political and financial predictions and forecasts have been proved correct.)

There have always been doomsday scenarios. The coming ice age. The global birth explosion. AIDS. Bird flu. The millennium bug. Politicians love threatening us. It is the easiest way to keep us frightened and easy to control.

But this book doesn't contain a prediction. This book is not simply my forecast for the future. This book is firmly based upon the facts we know.

And the consequences may be difficult to comprehend but they aren't difficult to see.

We like to think we are richer and more advanced than previous civilisations because we are more intelligent and work harder.
In our hubris we believe we have made progress through our talents and our wisdom.

Not true I'm afraid.

We have progressed, and become richer, because we discovered the temporary joy of fossil fuels.

`Energy is eternal delight,' wrote William Blake in `The Voice of the Devil'. As we have become richer so the earth's population has increased. In 1820 the human population hit one billion. Today it is six billion. Around five billion people alive today wouldn't be here if we hadn't discovered how to release the energy in fossil fuels.

Fossil fuels are the lottery win that have made us rich; a subsidy from nature which, instead of cherishing and using with respect, we have greedily plundered without regard for the survival of future generations.

Having discovered the joys of fossil fuel we bought them and stole them from poorer parts of the world when we had used up our own supplies. Some countries, such as the United States of America, have ruthlessly enriched themselves, and have risen to global power, at the expense of other nations.

Before we discovered how to liberate and use fossil fuels we had relied on manpower and horsepower for energy. Planting and harvesting crops could only be done as quickly as a man and a horse could move. Making things (houses, clothes, furniture etc) took one man a long time unless he found way to hire others to work with him or for him. This meant that for one man to increase his standard of living he had to find a way to exploit the labour of others. Kings and conquerors found this easy. They exploited men by threatening to kill them or imprison them if they didn't work. The end result was that for one man to increase his standard of living other men had to suffer - and their standard of living would fall.

And then, in the 19th century, along came the wonder of fossil fuels. And man discovered that the energy carried by these fuels could be exploited and turned into massively increased power, wealth and standard of living. Suddenly, everyone in the developed world could get rich. (Those in the rest of the world, where ironically much of the fuel was to be found, benefited only modestly.)

Man has always relied on natural resources.

The Romans and Greeks relied on marble. The British built their empire on ships made of oak and iron. Coal, dug out of the ground, provided a source of heat and energy.

But for the 20th century the one commodity resource that has fuelled the world more than any other has been oil.

Petroleum is the fossilised, compressed and baked remains of sea creatures and organisms which existed hundreds of millions of years ago. We use it to keep cars, lorries, buses, aeroplanes and trains moving. We use it to generate heat and to make electricity.

It was oil which took us out of the Steam Age and nothing in history has changed our lives in the way that oil has changed our lives.

Nothing.

The wheel, the plough, the spinning jenny, the steam engine, the telegraph, the internal combustion engine - all these things changed the way we live in significant ways.

But oil has changed our entire world.

It has changed the way we live our lives in every conceivable way. It has changed the way our economy can grow, it has changed the way we move about our planet, it has changed the way our factories operate, it has changed the way we live and the food we eat. It is no exaggeration to say that oil (and the other fossil fuels) have revolutionised our lives in every conceivable way.

It is, I suppose, possible to argue that it is the fossil fuels which have changed our world. It was, after all, the discovery that coal could be used as a fuel which led directly to the industrial revolution.

But oil is far more adaptable and far more useful than other fossil fuels. Oil has revolutionised agriculture, industry and domestic life. We speak of the Iron Age, the Stone Age and the Bronze Age and we must, therefore, also speak of the Oil Age. The difference is that to merit a description as an `Age' a civilisation usually lasts a few thousand years. The Oil Age will have lasted less than two hundred years by the time it is over. It's hardly surprising that energy and society expert Richard Heinberg describes us as living not in the Oil Age but in the Petroleum Interval or the Industrial Bubble.

You can't power motor cars with coal or uranium. No one has (as far as I know) seriously suggested putting nuclear powered or coal fired aeroplanes into the sky.

Oil has enabled us to leverage our knowledge and the power given to us by man's inventiveness. Oil has given us most of the things we cherish about our society. The construction of a single car requires at least 20 barrels of oil. Many modern drugs rely on oil. Mining and metal production rely on oil. And oil and other fossil fuels provide us with most of the energy we need to run our complex society.

And now the oil is running out. The world consumes three barrels of oil for every barrel discovered.

So, what will happen when the oil runs out? Cars, lorries, taxis, tractors, buses and trains will stop running. Planes will stop flying. Ships will stop sailing. There won't be any food (or anything else) in the supermarkets. There won't be any food because farmers won't be growing any food. Their tractors and combine harvesters will be parked, immobile, in their barns. Even if the farmers were growing food there would be no lorries to transport it to the shops.

Shops everywhere will be empty. They won't have anything to sell and they won't have any customers. No newspapers. No magazines. No new fashion handbags. No new computer games. No high definition broad screen digital television sets.

The lorry that collects your rubbish won't be running, although that won't matter much because you won't be able to buy anything and so you won't have any rubbish to throw away. Streetlights won't burn. Hospitals will have to close. Factories will shut their gates. Offices will close. Banks will shut. There won't be any more television programmes. You won't be able to recharge your mobile phone.

Within a generation five out of six people on the planet will be dead.

I'll repeat that.

Within a generation five out of six people on the planet will be dead.

This isn't the script for a horror movie.

It isn't fiction.

It's what is going to happen.

And it's already started.

The future I describe in this book isn't something that may happen.

It's something that will happen.

Peak oil - the point at which global oil production peaks - is, as I have already said, a more urgent, more critical and far more certain problem than global warming or terrorism.

In fact, although peak oil will exacerbate the terrorism problem (by encouraging greedy countries to start wars over the remaining oil) it will, to a large extent, solve the global warming problem. All the effort being spent at saving the planet from the effect of greenhouse gases is misspent. Global warming is not the primary threat to our present and future. When the oil runs out global warming won't get any worse.

Peak oil is enormously important to us all. It doesn't just mean that oil is going to get more expensive. It means an end to growth. It means an end to the sort of lifestyle we have become accustomed to. It means an end to industrial expansion.

The greens talk a great deal about alternatives such as wind power, solar power and wave power. But these alternatives cannot possible do any more than slow down the rate at which disaster approaches. None of these alternatives can come anywhere near replacing oil. Nor, as I will show, are biofuels an answer to anything except finding massive short term profits for special interest groups.

`There is no substitute for energy,' wrote E.F.Schumacher in 1973. `The whole edifice of modern society is built upon it…it is not just another commodity but the pre-condition of all commodities, a basic factor equal with air, water and earth.'

John Maynard Keynes, one of the greatest of all economists, pointed out that until very recently that had been no very great change in the quality of life of the average man living in a town or city.

Keynes estimated that the standard of living just about doubled between 2000 BC and the start of the 18th century. That's nearly 4,000 years of hardly any progress at all.

At the start of that time we had fire, the sail, the wheel, cooking utensils, banks, governments, religion, mathematics, ploughs, farming and language. And by the start of the 18th century that was still pretty well all we had.

And then in 1712 an Englishman called Thomas Newcomen built the first steam engine. The Industrial Revolution was started by a man from Devon.

Over the years that followed fossil fuels changed our world, giving us electricity, fertilizer, steel and plastic. The standard of living for the average man living in the average town or city doubled every few decades. And kept doubling for quite a while.

Newcomen's steam engine was originally built to pump water out of a coal mine.

The Industrial Revolution revolved entirely around fossil fuels. It was coal and oil which changed our economy from an agrarian, handicraft one to an economy dominated by industry and machine manufacture. It was the Industrial Revolution which led to the use of iron and steel, instead of wood, and, eventually, to the introduction of new energy sources such as electricity. It was the Industrial Revolution which led to the invention of new machines (such as the spinning jenny), the development of the factory system and to the development of the steam engine, the telegraph, the internal combustion engine and the jet engine. It was the factory system, a result of the Industrial Revolution, which led to the development of schools (so that there would be somewhere for children to go while their parents worked in the factories, and so that children would grow up accustomed to a day spent working) and terraced housing (so that workers could be accommodated close to the factories where they worked).

The Industrial Revolution resulted in changes in agriculture (tractors instead of horses), political changes (workers, now paying tax, wanting votes) and enormous social changes.

The Industrial Revolution was largely confined to England, and then Britain, until 1830. It then spread to France before reaching Germany and, eventually, the USA. Now England's great Revolution has spread to

China, India and the rest of Asia. Everywhere that the Industrial Revolution went it was built upon a supply of fossil fuels.

Coal was the first fossil fuel to change our lives.

Before mankind discovered the benefits of coal our sources of energy were food and wood. Energy depended entirely on stuff we could grow - using our own muscles to do the digging and the sowing.

When men started digging coal out of the ground they started using energy sources that were already in existence - and had been formed generations before. Coal, oil and other fossil fuels are just what the name says: fossil fuels. They are created when ancient bits of matter are steadily crushed by billions of tons of rock. It takes millions of years for fossil fuels to form.

Coal was being burnt for heating and cooking in China 4,000 years ago. It was used in mediaeval Europe too, though it didn't overtake wood as a fuel because it had to be mined and transported - both of which required a good deal of effort and energy.

By the early 17th century English manufacturers producing iron and steel discovered that the higher temperatures possible with coal made it easy to smelt iron and work with metal.

But it was still difficult to get coal out of the ground. The biggest problem was that water tended to accumulate at the bottom of the mine shafts. And, as I've already described, in 1712 this problem was solved when Samuel Newcomen invented a simple steam engine specifically to pump water out of coal mines. And so, slowly, the industrial age was born out of the rediscovery of coal.

In 1803 an English engineer called Richard Trevithick used the improvements devised by James Watt and installed a steam engine on a carriage, intending it for use on the roads. Unfortunately, roads hadn't yet been invented and the steam carriage wasn't much use until

George Stephenson (another Englishman) put the steam locomotive on rails. Not surprisingly, the rails he used were similar to those used in the tramways in coal mines.

Things moved swiftly after that. In the 1790s an English engineer lit his factory with gaslights. In 1804 gas lighting was installed on the streets of London. By 1840 steam engines were being used on ships. And in 1854 coal-tar dyes were discovered and the chemical industry was born.

In 1800 the annual world coal output was 15 million tons. By 1900 the annual world coal output was 700 million tons and coal had transformed the world. The 19th century was the Coal Age. (Or, perhaps, the Coal Interlude.)

From that point on the world's energy would be derived not from renewable resources (human and equine muscle strength) but from a source of energy that, once gone, could not be replaced.

As machines became more widespread during the 19th century so there was a need for oils to lubricate them. Whale oil, animal fats and vegetable oil were all used. Whale oil was also used as a fuel for lamps. Using so much whale oil meant that whales were hunted pretty well to extinction.

Petroleum oil had been used since the 7th century when Byzantine Emperor Constantine IV fixed flame throwers to the prows of his ships and the walls of the city when defending Constantinople. The flames were created using a mixture of naphtha, quicklime and sulphur which was known as `Greek Fire'. Ironically, two Saracen fleets were destroyed with Greek Fire. Oil had also been used as a lubricant and as a medicament.

However, the only petroleum oil available was the stuff that seeped to the surface of the earth.

The first commercial oil well was drilled in the mid 19th century in America and from then on oil was used increasingly as a lubricant and as a lamp oil.

Ruthless American oilman Rockefeller used industrial espionage, predatory pricing and a variety of other dirty tricks to take over foreign oil companies and by 1865 had very nearly obtained a worldwide monopoly on petroleum supplies.

By the early 20th century oil was being used as a fuel for factories, trains and ships and oil burning furnaces were becoming common.

Oil has enormous benefits: it is easy to transport, it's full of energy and it can be refined into a variety of different fuels (diesel, petrol, kerosene) which can be used in many different ways.

Natural gas, often found alongside oil, was also brought into use for street lighting.

And then came electricity.

The first electric generator was invented in London in 1834 though, as with trains, cars and aeroplanes (all of which were invented in Europe) it was first commercialised in America.

Electricity isn't a fossil fuel, of course. It doesn't occur naturally in great seams in the ground. Coal, oil, gas, uranium or some other source of energy have to be converted into electricity.

It is the fossil fuels which enable us to obtain electricity cheaply and easily. Without them, and other sources of renewable energy such as water (either in the form of rivers or the sea) and the wind, we would have to create electricity by using muscle power and that would prove tedious and exhausting.

The special powers of electricity have enabled us to use it to power a vast number of devices - ranging from industrial machinery to television sets. It's clean and enormously convenient.

The problem with electricity is that as a carrier of energy it is extremely inefficient all along the line - from the initial energy source right through to the final point of use.

However, as long as we are getting our energy cheaply from fossil fuels the wastefulness of electricity doesn't matter much. In his book The Party's Over, Richard Heinberg reckons that: `At current prices, an amount of electricity equivalent to the energy expended by a person who works all day, thereby burning 1,000 calories worth of food, can be bought for less than 25 cents.'

Much of our modern lifestyle is dependent upon a steady supply of electricity.

And the production of electricity is pretty much dependent upon fossil fuels. Only a tiny proportion of the electricity the world uses is derived from renewable sources.

There has, of course, been a price to pay for our reliance on fossil fuels.

Our burning of fossil fuels is now held responsible for global warming - and all the horrors that global warming may bring.

When the Industrial Revolution started, at the beginning of the 18th century, the earth's atmosphere contained 275 parts per million of carbon dioxide. It now contains over 380 parts per million. And the figure is rising fast. It is, say the experts, the highest the planet has seen for millions of years.   Climatologists claim that the earth's temperature will rise by another four or five degrees (at least) before the 21st century is out. This would make the earth warmer than it has been since before primates first appeared.

As the world's atmosphere warms it will generate more frequent heat waves, longer droughts, more severe storms and more intense downpours. There will be flooding in some areas and droughts in others.

Global warming is a serious problem.

But as problems go it isn't in the same league as peak oil.

There are two reasons for this.

First, the oil is running out. So the Kyoto Protocol and other treaties are unnecessary. The Americans will have to cut down their oil consumption when the stuff runs out. The end of the oil is a primary problem. Global warming is a secondary problem.

Second, society as we know it is going to be destroyed by peak oil. How more serious than that can anything be?

The big problem is that although we have plenty of coal left we are reaching the bottom of the barrel as far as oil is concerned.

The benefits of fossil fuel are extraordinary.

Without fossil fuel it would take five people working continuously to create enough power to keep a 150 watt bulb burning. A motor car on the motorway uses up the sort of energy that might be produced by 2,000 people.

In `The Party's Over' Richard Heinberg explains: `If we were to add together the power of all the fuel-fed machines that we rely on to light and heat our homes, transport us, and otherwise keep us in the style to which we have become accustomed, and then compare that total with the amount of power that can be generated by the human body, we would find that each American has the equivalent of over 150 `energy slaves' working for us 24 hours each day. In energy terms, each middle-class American is living a lifestyle so lavish as to make nearly any sultan or potentate in history swoon with envy.'

For the last 100 years or so we have had the joy of using a virtually free energy source. All we had to do was take it out of the ground. The energy in a gallon of petrol is approximately the same as the energy expended by a man working hard for a month. In America petrol is so cheap that a man receiving a minimum wage can buy a gallon of the stuff for the price of about 20 minutes work. As a result the Americans use up petrol faster than they use water. If everyone on the planet consumed oil at the rate of the average American the world would need 450 million barrels of oil a day (compared to the 86 million barrels a day needed at the moment). We would have probably run out of oil already.

Anyone in a western nation has, for years, effectively been able to rely on having hundreds of energy slaves working for him (or her).

Before oil a man would need to expend great personal energy to travel thirty miles. With oil such a journey becomes a trivial adventure.

Finding oil - with its latent energy - has been the equivalent of a mass lottery win. Coal is a useful fuel (although it is extremely dirty and a major cause of pollution) but it isn't anywhere near as versatile as oil. There aren't many things you can do with coal that you can't do with oil but there are a lot of things you can do with oil that you can't do with coal.

How many people do you see driving around in coal fired cars? How many coal fired aeroplanes are there?

But instead of using our find to improve our world, and to eradicate poverty and hunger around the world (as we could so easily have done) we have used our find to help us build private aeroplanes, luxury yachts, space rockets, dish washers and petrol driven lawnmowers. We have invented a thousand ways to use up the energy we have. Populations have expanded and governments have grown fat on the taxes they have imposed on the new millions. We have used and abused the planets resources as though they were limitless. We have damaged and polluted the planet which has given us so much.

And we have ignored the reality of our increasing dependency on a substance that is running out.

We have learned to take the benefits of fossil fuel for granted.

But the fossil fuels won't be around for much longer.

For the last century or so we've been very lucky.

Now our luck is running out.

And when the fossil fuels (particularly oil) disappear the earth will only be able to feed and provide shelter and warmth for a much smaller global population. Farmers will no longer be able to use fertilisers and tractors. Combine harvesters and lorries will stand and rot. Farms will produce what can be cultivated and harvested by the labour of men and horses.

The end result will be that there will be six billion people living on a planet capable of sustaining one billion people.

Here's what Richard Heinberg, the author of `The Party's Over' says: `When the flow of fuels begins to diminish, everyone might actually be worse off than they would have been had those fuels never been discovered because our pre-industrial survival skills will have been lost and there will be an intense competition for food and water among members of the now unsupportable population.'

And what are we doing to prepare ourselves for the disaster ahead?

Nothing.

Our politicians believe that someone, somewhere will find another oil field and solve all our problems.

They believe that someone, somewhere will discover something else we can use to replace the disappearing oil.

But there isn't any more oil.

And there isn't anything out there with which to replace it.   Our luck is fast running out.

Of course, it is possible to argue that being lucky isn't always an entirely good thing.

Great wealth can have a terrible effect on the person (or community) acquiring it.

Lottery winners, taken from poverty to multimillionaire status in a moment, frequently report that their lives have been ruined by their good fortune. Communities can be devastated for years by the arrival of unexpected wealth. (As may happen when a village which has been at the centre of a disaster is overwhelmed with financial gifts far in excess of what is needed for the repair work.)

Cataclysmic wealth can affect an individual or a community or an institution or a nation in bizarre and utterly disastrous ways. New habits and expectations are quickly formed. Previously established ideas are discarded in the belief that they no longer matter. Usefulness is no longer a factor. Waste seems to become a way of life. Excess is suddenly laudable.

Our discovery of the energy treasure trove hidden just beneath the surface of the earth has affected our society in an almost wholly reprehensible way.

Instead of using our wealth to do good things (for example, making a serious attempt to eradicate poverty and hunger in the nations from which we have helped ourselves to so many natural resources) we have used our newfound wealth to buy fast cars and faster aeroplanes and to build ourselves vast mansions, skyscrapers and absurd and ugly temples to our passing glory.

Slowly, as we begin to recognise that our wealth is disappearing, that we have spent too much and too quickly and that we have failed to provide for our future, so our desperation increases.

The desperation to `stay lucky' has triggered a good deal of violence.

American leaders have for some decades been only too aware that their nation's continued success and status as a superpower depends upon seizing the world's diminishing supplies of energy.

You can't drive tanks or fly jets and bombers without oil. You need vast amounts of fuel to force aircraft carriers through the sea. Even nuclear powered submarines need fuel. Without fuel nothing moves.

But those same leaders who recognise the need to grab whatever fuel they can are too frightened of the voters to tell the truth; too frightened of unpopularity to shout the warning about diminishing resources, the need to hold back, the desperate need to stop wasting, the too-late need to save a little for tomorrow.

It is, it seems, with nations as it is with individuals.

When the money starts to run out the lottery winner becomes desperate. He starts buy more lottery tickets in the hope that lightning will strike twice. He starts blaming other people for his impending return to poverty. He lives on hope. It doesn't occur to him that his future could still be secure if only he would stop spending, take stock and be a little more prudent. He becomes angry and resentful. He may take to gambling in a big way; putting great chunks of whatever remains on horses or numbers on the roulette wheel.

The young heir who has blown his legacy on smart cars, cocaine and nightclubs will look around in bewilderment and anger as he realises that his fortune has disappeared. He will borrow from his friends. He will become an inveterate sponger.

The one thing they have in common is their desperate desire to maintain their standard of living. They do not want things to change. They are accustomed to wealth and to them it does not seem fair that they should now be deprived of what they have come to see as their rights.

Societies are the same.

As I have explained in previous books large organisations develop lives of their own. The large corporation is not controlled by its shareholders or its directors. The corporation itself dictates what happens next according to its own needs. Similarly, nations develop lives and characters of their own.

All have one thing in common: an insane desperation to maintain things as they are.   And all end in the same way: disintegration, violence, despair, sadness and collapse.

Sometimes, it seems, there isn't as much difference between good luck and bad luck as we might like to think there is.

Writing in Scientific American magazine in March 1998, Colin J. Campbell and Jean Laherrere wrote: `From an economic perspective, when the world runs completely out of oil is...not directly relevant: what matters is when production begins to taper off. Beyond that point, prices will rise unless demand declines commensurately. Using several different techniques to estimate the current reserves of conventional oil and the amount still left to be discovered, we conclude that the decline will begin before 2010.'

At least 85% of the energy consumed by the world comes from fossil fuels.

That means that when the fossil fuels run out we will lose 85% of our energy.

This means that our world is about to change beyond recognition.

Without oil we might as well start living on a different planet.

Of course, no one knows for certain how much oil is left, how difficult it will be to extract or how fast the demand for oil will grow. We   will   only really know when the oil started to run out some years after it started to run out. But the available evidence firmly suggests that if it hasn't already started then it will do so within before the end of this decade.

The oil isn't going to run out immediately, of course.

There is still a lot of oil left in the ground.

But the amount we will be able to get out of the ground will now diminish every year. Whatever oil reserves there are will be increasingly difficult to extract.

The shortage will be made worse by the fact that the demand for oil is rising every year.

And it is this gap between what is available and what is required which will lead to the first, early problems.

How much difference will it make when the oil shortfall reaches 10%?

Well, the price will soar much more than 10% of course.

And there will, inevitably, be a lot of people who get less oil than they need.

The world at large has no idea what is about to happen. Believe it or not, oil is cheap at the moment. But the days of cheap oil are almost over. I wonder how angry motorists and lorry drivers will be when the price per gallon hits £30 and keeps moving upwards?

That's when the problems will really begin.

This year? Next year?

I don't know.

But it will be sometime soon.

And once the problems start they will get worse day by day, week by week, month by month and year by year.

Meanwhile, every country in the world wants to capture some of the remaining fuel. America has started a war in just about every country in the world where it might conceivably hope to find some oil. But America is not the only powerful country hunting for energy supplies.

The Chinese Government understands the importance of securing energy supplies. (The west accuses China of bankrolling genocide in Darfur to obtain oil but this, of course, is grotesquely hypocritical, considering the way America and Britain have abused Middle Eastern oil producers.) China, once an oil exporter but now an importer has, of course, also done some clever deals with Iran.

And China's immediate Asian competitors are competing for the same energy supplies. Japan and South Korea have both agreed to store foreign oil (from Saudi Arabia) in exchange for preferential access in a crisis. Japanese power utilities have agreed to buy uranium from Kazakhstan and will get as much as a third of the fuel they need for Japanese power stations from Kazakhstan by 2010.

Everywhere you look countries are doing energy deals with one another.

China, now a net importer of coal for the first time in history, is also looking for coal. (Around 80% of China's electricity comes from coal fired plants - which are largely responsible for the filthy, smoky air now choking the lungs of Chinese citizens.)

As Stephen Leeb points out in `The Coming Economic Collapse': `We have built our modern civilisation on the premise of unending growth - growth that needs energy. We have built a complex civilisation that requires increasingly larger amounts of energy to maintain itself. What happens if growth is no longer possible? What if...we start to run out of the resources needed to build bigger cities? What if even maintaining the cities we have today becomes too expensive?'

But once growth stops, all sorts of really terrible things will happen.

Our entire global economy is built upon the need for growth. Our financial system is built upon fiat currencies. We rely on bits of paper to represent wealth. Currencies represent the wealth of the issuing nation. Stocks and bonds represent the values of individual companies. Options, futures and so on are simply certificates which represent something of value. None of these things are of any value in themselves. A twenty pound note has no real, intrinsic value and depends entirely upon the honesty, integrity and promises of the government which printed it and put it into circulation. A share certificate only has value because it represents part ownership of a company.

The problem is that in our modern world these bits of paper represent a debt. In all these cases the purchaser assumes that the issuer will grow. Governments and companies borrow vast amounts of money in order to keep going. The only way they can ever pay back their debts is by growing.

If growth stops then wealth doesn't grow any more.

And the bits of paper - the currencies and the share certificates - will rapidly lose their value.

Our society is more complex than any other society in history. We have become dependent on one another in a way never seen before.

(There is some irony in this because our society has also broken down in many ways. The Government has deliberately done all it can to destroy marriage and the family unit. We have been encouraged to be fearful of our neighbours and to distrust strangers. This fear and distrust begins at an early age with small children being repeatedly warned that they must not talk to strangers or take sweets from them.)

We depend upon the shops (or, more likely, the supermarkets) for our food. We depend upon the electricity, gas and water companies for our electricity, gas and water. We depend on others for all our health care and for our protection. When the television set breaks down we call a television specialist. When the washing machine breaks down we call a washing machine specialist. When the dishwasher breaks down we call a dishwasher specialist. When the car breaks down we need a mechanic. When the computer breaks down we need a computer expert (or a shop selling new computers). When the telephone breaks down we are done for.

As individuals we have become specialists and we have learned to acquire specific skills which help us fit in with the needs of our complex society.

And everywhere you look you will see that our society depends on oil.

The washing machine is built using vast amounts of oil. It won't work without electricity, probably created from oil. And when it breaks down the repairman drives to your home in a car using petrol.

Our supercomplex society is built on oil.

And now the oil is running out.

And just as the Roman Empire collapsed, and the Greek and Egyptian civilisations virtually died away, so our civilisation can die.

Joseph Tainter, author of The Collapse of Complex Societies, has concluded that complex societies tend to collapse because their strategies for energy capture are subject to the law of diminishing returns.

The Romans tried to capture energy by conquering new lands and acquiring slaves.

We have captured energy by taking fossil fuels out of the ground.

The Roman Empire collapsed because the Romans could no longer manage such a huge empire effectively. The cost of looking after all that land and all those slaves eventually grew greater than the benefits. When the Romans first invaded a new country there were immediate benefits: booty to be taken. But, over time, as the benefits of the invasion declined, so the costs increased. There were garrisons to be maintained and administrators to pay. To find the funds for all this taxes had to be raised. And so eventually the productive capacity of the population declined. There were too many people collecting taxes, making rules and administering and there weren't enough people growing food, making things and carrying spears.

If our civilisation collapses it will be because we grew to depend on a finite source of energy while at the same time our society had become so complex that we spent huge amounts of our `energy budget' on maintaining nonproductive organisations, institutions and bureaucracies.

In simpler societies there were only a few dozen different occupations. In our society there are thousands of ways for men and women to earn a living. We have become intensely specialised and dependent upon one another. But, most dangerously of all, we have built up hugely complex management structures.

`However much we like to think of ourselves as something special in world history,' wrote Joseph Tainter, `industrial societies are subject to the same principles that caused earlier societies to collapse.'

The theory of peak oil (which I explain more fully a little later in this book) effectively states that once half the oil is removed from an oil field the amount of oil which can be removed will decline, often quite rapidly.

Britain has already passed her point of peak oil production and her own oil supplies are running out. Oil production by nations outside OPEC peaked at the end of the 20th century and is now in decline. According to the peak oil theory our consumption of oil has (or will soon) outstrip the production from existing reserves and the discovery of new reserves. Whenever that point is reached we start to deplete existing reserves.

America and Britain now depends largely on the Arab members of OPEC for their future oil supplies. These, of course, are the same Arab states that these Governments have done their very best to annoy.

The end result will, of course, be that the Arab nations will be reluctant to let these two countries have any of the increasingly rare and expensive oil when the stuff really does start to run out.

The USA passed its own peak oil point over thirty years ago and has since then been totally dependent on whatever oil it can buy from other countries.

Unfortunately for the Americans, of course, the USA is not the most popular nation on earth as far as the Arabs are concerned. (Britain's popularity is, thanks to the war against Iraq, now inextricably linked to America's.)

Nor are America and Britain doing too well in their relations with the other big oil suppliers: Russia, Venezuela and Nigeria.

Even if the oil wasn't running out the future would be bleak.

It's hardly surprising that America has been trying to grab as much oil as it can.

Meanwhile, the world's consumption of oil is rocketing. The Americans use more and more oil every day. They are still the greediest oil consumers. Their consumption has increased by just under three million barrels a day in the last decade. American Government subsidies mean that the stuff is still cheap enough for people to drive huge cars and trucks rather than smaller, more efficient vehicles.

It isn't just America which is using more oil than ever, of course.

China and India are growing. Their citizens are buying cars and learning the joys of international flight travel.

One in six people in the world live in China. They all want cars. Over a billion of them. If car ownership in China were to rise just to American levels there would be 650 million cars on Chinese roads. that's more than all the cars in the world today.   People who, ten years ago, could hardly afford bicycles are now getting picky about the size of the car they buy. China, now the world's second biggest oil importer, is currently increasing its oil imports by 30% a year. How much longer will it before China catches up with America? In 2003 Shanghai banned bicycles from most of its main streets to make more room for automobiles. China already has 16 of the 20 most polluted cities in the world. In Bangkok the average motorist spends the equivalent of 40 days a year stuck in traffic. Half the new houses and apartments being built in China are equipped with air conditioning. That pushes up the electricity consumption dramatically.

In America even the poor keep their television sets on day and night, fly around in aeroplanes and believe air conditioning and central heating are a basic right.

In Europe new cut rate airlines are buying new planes and opening up new routes faster than ever. Every time a plane takes off, and flies a bunch of holiday makers to Marbella or Miami, it uses up another chunk of the world's depleting supplies.

Economists and politicians are agreed that there is nothing for us to worry about. Politicians don't want to tell us really bad news for which there are no solutions. They don't like dealing with crises before they have happened. And they certainly don't like crises and problems they can't blame on someone else. Economists claim it doesn't matter if we run out of one fossil fuel.

They argue that if we run short of something the world will pay for someone to find something else we can use.

Politicians like this theory. It makes them feel comforted.

These heads-in-the-sand deniers of the truth are aided in their deceit by the oil-producing states which constantly produce comforting sounds, designed to reassure us.

But independent petroleum geologists (the people who know) are worried. And their fears have been well documented in the relevant scientific literature for some time.

They say that the oil is running out and that it is not going to be possible to find a substitute. The chances are that well within your lifetime (unless you've considerably over eighty years old) the world we know will change dramatically and permanently. Anyone under fifty, with a normal life expectation, will live to see a world almost unrecognisable from the one they grew up in.

We are often reassured that scientists always find solutions to our problems - especially when there is a financial incentive to do so.

We are told that scientists are constantly making new breakthroughs.

But is this really true?

I don't think so.

On the contrary, I believe that there has been very little real scientific or technological progress in recent decades.

Look around. Most of the brilliant devices we use were first put on the market many years ago. They have been improved and miniaturised. But that's all.

It is, of course, perfectly true that information technologies are doubling in power every year. Doubling every year means multiplying by 1,000 every ten years. And, on this basis, scientists talk merrily about the advances in computing power, revolutions in genetics, robotics and nanotechnology as though these inventions are going to improve our lives and solve our problems.

But the truth is that none of these things will really help with the big problem. We have more than enough information technology.

It's oil we need.

Improvements in information technology are not going to help us deal with peak oil.

The futurologists have talked about all sorts of things that they claimed would change our lives. None of them have arrived.

But look at the great inventions which changed our lives:

The first motor cars were built in the mid 19th century.

The internal combustion engine was built in 1885.

The telephone was invented in 1876.

The camera was invented in the early 19th century.

The first antibiotics were discovered in 1928.

The first open heart surgery was performed in 1896.

The atom was first split in 1919.

The first engine powered aircraft were built by English and French inventors in the 1840s.

The computer was invented in the late 1940s.

The fax machine was invented in the 19th century.

The television was invented in the first decade of the 20th century.

You could fly across the Atlantic in six hours in the 1960s. It takes the same time today. (But it's less comfortable.)

The truth is that there have been no real breakthroughs for years. And none of the minor inventions which have been patented have improved the world's food supply, offered a realistic solution to our growing energy problems, or improved the world's water supply.

The so-called breakthroughs which have been made have simply been minor improvements.

Cameras and telephones have become smaller - but little more effective than their predecessors. Computers have become faster and more portable - but the changes haven't been life altering.

We like to think we're cleverer than any other generation. But the evidence does not support that theory. We like to think we will be able to `think' our way out of all our problems. The evidence suggests otherwise.

Politicians (aided by a few friendly experts) claim that the oil will last for ever and that there is nothing to worry about.

`What is not in doubt,' wrote Andrew Alexander in the Daily Mail on 16th September 2005, `is that there is lashings of the stuff around, consume it as we may....'. Mr Alexander didn't say where this oil was to be found (though I have no doubt the oil companies would have been grateful for guidance).

In the spring of 1999 The Economist (which, in my view, has an unparalleled knack for getting big stories wrong and for misinterpreting the evidence on a wide variety of issues) produced a cover story called `Drowning in Oil' which suggested that Saudia Arabia had decided to flood the world with enough extra oil to take oil prices down to five dollars a barrel, and to keep oil at this low price for at least five years. Just a few days after the publication of this story the Saudi Arabians, together with other countries, cut their production of oil. The price of a barrel of oil went up threefold in 18 months.

Oil companies will, say the optimists, simply find some more when the existing oil fields run dry. They are not, it seems, aware that since 2001 the cost of looking for oil has exceeded the value of the oil discovered. (And, of course, the energy expended in looking for that oil has almost certainly exceeded the energy obtained.)

Writing in `The Skeptical Environmentalist' in 2001 Bjorn Lomborg argued that: `It is rather odd that anyone could have thought that known resources pretty much represented what was left, and therefore predicted dire problems when these had run out. It is like glancing in my refrigerator and saying: `Oh, you've only got food for three days. In four days you will die of starvation.' But in two days I will go to the supermarket and buy more food. The point is that oil will come not only from the sources we already know, but also from many sources of which we do not yet know.'

Once again, if Mr Lomborg knows, or even suspects he knows, where oil might be found he should share the information with the world's oil companies. They would doubtless pay him well for the information.

Writing in The Prize, Daniel Yergin says: `The ultimate amount (of energy) available to us is determined both by economics and technology.' He goes on to add, with commendable optimism: `If you pay smart people enough money, they'll figure out all sorts of ways to get the oil you need.'

Sure.

From Mars maybe? Huge intergalactic tankers floating through universe?

And if the oil does happen to run out, say the optimists, hedging their bets, someone will find an alternative. Man's ingenuity will save the day.

An alternative to oil will be found, they say, and then all will be well. Things, they say, can carry on as they are for ever.

But many thousands of people have spent many years looking for real alternatives without success. And even if someone found a good, alternative source of energy political will would be needed if we were to exploit the idea to best advantage. There is clearly no such political will.

Most importantly, of all, however this argument completely fails to acknowledge the main problem: fossil fuels were nature's gift to us; they were effectively a one-time only lottery win. Fossil fuels gave us free energy - available only for the cost of extracting it. We will not match that generosity until and unless we discover either another variety of fossil fuel or the perpetual motion machine. And inventors have been struggling to create a perpetual motion machine for thousands of years.

There are those who say it doesn't matter that the oil is running out.

Nobel prize winning economist Robert Solow was quoted as saying that: `...the world can, in effect, get along without natural resources.'

Really.

A world without modern agriculture, without cars, without aeroplanes, without ships, without heating, without much of our electricity, without industry and..well, do I need to go on to demonstrate the dangerous absurdity of this overconfidence?

As unpalatable as the fact may be, there is only so much oil on our planet.

Billions of pounds have been spent looking for it everywhere. The oil exploration industry is, next to the military, the world's biggest consumer of computer technology.

And it is clear that there isn't much more oil to find.

The few new oil fields which are discovered are small and costly to use.

American oilmen have been struggling to find new oil for decades. But they have failed to prevent the decline in American oil production. As Stephen Leeb puts it in `The Coming Economic Collapse': `Money and brains cannot repeal basic geological laws'.

Moreover, we cannot invent an oil substitute any more than we invented oil. We may be able to invent new sources of power, but they will be limited by the laws of nature. Oil, as a source of energy, never had to obey the laws of nature because it was a product of nature.

Huge multinational companies have spent the last few decades desperately searching for a replacement for oil.

And they have failed.

Environmentalists claim that we can do without oil by harnessing the power of the sun, the power of the sea and the power of the wind.

Nice idea.

But it won't work.

These all collect daily energy. They sound a good idea. But they won't replace fossil fuels.

Fossil fuels offered us `free energy'. Mankind's first and only really free lunch. All we had to do was dig them up. Centuries and centuries of accumulated power.

Sunshine, wind and wave power are a nice idea. But, for reasons which I will explain later in this book, they are no replacement for oil.

It's not working out quite as the optimists would like.

There is no new wonder replacement for oil.

There are a number of reasons why our leaders are sticking their heads in the sand and refusing to do anything about the biggest problem our society has ever faced.

First, most of our leaders aren't terribly bright but are crafty and mendacious. They are concerned solely with their own survival.

We have created a society which, for various complex reasons now tends to select crooks and conmen as leaders. And we have allowed them to create a self-preserving oligarchy.

Second, although politicians love telling us about things that will frighten us (there is nothing easier to manipulate than a frightened electorate) they know that there is little political capital to be gained by telling the truth when the truth involves genuine short term pain for everyone. Our politicians think only about their survival. Why should they tell us something we don't want to hear?

Third, many of the things that need doing (such as strengthening families and small communities) are an anathema to our politicians. The Government has realised that its power is increased when the family is weak.

Fourth, modern politicians are herd animals. They don't like doing anything that doesn't fit in with their peers are doing. Modern politicians rely not on integrity or honour but on something called `group think'.`We need an energy bill that encourages consumption,' said George W. Bush in 2002.

Some years ago a psychologist called Solomon Asch performed an experiment to find out whether people's tendency to agree with their peers was more powerful than their tendency towards rational and independent thought.

Asch collected a dozen students and told them to look at a drawing of three lines. He told them to decide which line was the longest. The right answer was obvious.

But Asch had told eleven of the twelve students to lie and say that the medium length line was the longest. The aim was to see whether the twelfth student would stick to his guns (and point to the line that really was the longest) or would capitulate and go along with his lying chums. The twelfth student was the only real object of the experiment.

Amazingly, in 70% of the tests Asch did the twelfth student gave in and said that the medium length line was the longest.   Asch's conclusion was that most people will follow the crowd even when the crowd is clearly wrong.

Later, one of Asch's students, Stanley Milgram, performed another, even more famous experiment. His subjects (all ordinary people) were told by an authority figure (a man in a white coat who looked and sounded like a scientist) that they had to inflict pain on another person. And 65% of them did what they were told. Even though they thought they were inflicting serious and possibly lethal pain on a fellow human being for no good reason whatsoever they did it.

Milgram showed that most people trust and obey those in authority so much that they will abandon all their personal standards, all sense of personal responsibility, all feelings about what is right or wrong, and blindly follow orders.

This is `group think'.

A third psychologist, Milton Rokeach, subsequently argued that there are two distinct types of people.

First, there are those who like hierarchies and who are happy giving or taking orders. These people tend to have closed minds and to resist new ideas. Because of the way our society is structured these are the people who obtain positions of power.   Second, there are those who are instinctively opposed to authority and who think freely and creatively.

It doesn't take much imagination to realise that most (if not all) of our leading politicians are firmly in the first group. They have a narrow view and tend to ignore problems until it's too late.

Virtually every leading politician we've had since Winston Churchill has been unimaginative and hierarchical.

Politicians make millions follow them not by courageous, strong, honest leadership but by `group think'.

Arrogance, ambition and closed mindedness (the prominent traits among modern politicians) lead to an inability to make clear, accurate moral judgements.

Hans Blix, the chief United Nations weapons inspector, accused American of relying on `faith-based intelligence' in claiming that Iraq had weapons of mass destruction.

The American leadership wanted to believe that Iraq had weapons of mass destruction. It was a convenient belief. So, no one questioned it. And the CIA, and other organisations, rejected any information which didn't fit their requirements.

(In fact, as we now know, the American leadership was inspired by a desire to acquire Iraq's oil. That was their driving force. But the willingness of the intelligence community to fit in with their preconceived and pragmatic ideas was a perfect example of `groupthink'.)

`Continuing to increase our dependency on petroleum consumption is clearly a suicidal course of action,' wrote Paul Ehrlich in 1974. `The only intelligent alternative is to begin reducing energy consumption and finding alternative energy sources to substitute for petroleum.'

Our leaders aren't doing anything about the disappearing oil.

And they won't do anything until it's far too late. They will respond to the crisis when it becomes a crisis. And when their actions will be meaningless and irrelevant.

They will respond when the only thing they can do is to put troops and police on the streets to arrest and shoot rioters.

(There is, of course, a line of thought which suggests that our leaders are deliberately allowing the crisis to develop so that they can increase their hold on us. Maybe. It's not a theory to which I subscribe. Only because I don't think they are that clever.)

A world without oil will be a world without cars, aeroplanes, combine harvesters, ships, motorbikes, fertilisers, plastic, washing machines, television sets, computer games, pesticides and a hundred thousand other things we take for granted.

It will be a world where heating your home will become prohibitively expensive and where air conditioning will be something only the superrich can enjoy.

And a world with less oil will be almost as bad as a world with no oil because once it becomes clear that the oil is running out then governments worldwide will start to stockpile the stuff. And just a 10% cut in the oil supply will mean a dramatic change in our way of life.

Fossil fuels in general, but oil in particular, gave us a free ride throughout the twentieth century. They enabled us to do less work but produce more. They enabled us to make enormous progress and to acquire a taste for luxury and a life of waste. All we had to was find the oil, coal and gas and then take it out of the ground. We didn't have to do anything to create the energy. We've been spoilt. We have for over a century lived in luxury; enjoying the windfall the planet has given just a few select generations.

And now the party is almost over.

Our lives are about to be changed for ever. Nothing we take for granted will be the same.

Everything about our world will inevitably change. And the changes will be dramatic and irreversible.

No other threat in our history comes close to this one.

Countries with weak farming industries will be in serious trouble when the oil runs out - and will find it particularly difficult to survive. Britain will suffer more than almost any other country on the planet. If you're an island and you've destroyed your farming industry, and there is no oil for planes or ships, just where do you get your food from?

And, of course, its Government's illegal invasion of Iraq has made Britain one of the two most hated nations on earth. In addition to making Britain a terrorist target the Government has ensured that when the oil is running out the British will not be the people getting a share of whatever is left.

Of all the developed countries in the world Britain is the least well positioned to cope with the end of oil.

The citizens of every country will have to cope with enormous change and will suffer terrible hardships. The citizens of Britain will have to cope with bigger changes than most and will suffer greater hardships.

All countries are heading for compulsory downsizing. Large seemingly invincible corporations will go bankrupt (taking with them the pensions of millions). Unemployment will hit unprecedented levels. Interest rates will almost certainly rocket. Many currencies will not survive in their present form.

The coming change is not voluntary. It's not optional. And it's not temporary. It's going to happen; whether you or I like it or not.

Every aspect of our life will be permanently changed. Even the administration of the nation will alter. The European Union will collapse, national governments will decline in importance and regional administrations will shrink dramatically. The Government's income will fall dramatically (as companies collapse and unemployment soars so the tax take will shrink dramatically). Those who thought their employment was safe will find themselves joining the millions officially listed as unemployed. There will, however, be little point in bothering to register as unemployed. All benefits and pensions paid by the Government will halt or shrink to pitiful amounts. Former civil servants and local authority employees who thought that their pensions were secure will be penniless when the Government and local authorities are unable to meet their financial obligations.

`In 1859 the human race discovered a huge treasure chest in its basement,' wrote Kenneth E.Boulding in 1978. `This was oil and gas, a fantastically cheap and easily available source of energy. We did, or at least some of us did, what anybody does who discovers a treasure in the basement - live it up, and we have been spending this treasure with great enjoyment.'

Twentieth century industrial civilisation is founded on the consumption of finite energy resources. Oil, gas and coal have given us a free ride for decades. Now they're running out. We have built a lifestyle based on consuming vast quantities of fossil fuel. Agriculture, transport, communication, industry, society - everything we rely on, everything we do is dependent upon a continuing, uninterrupted supply of cheap fuel.

But the fuels are already running out. And the clock is ticking.

As they run out the competition to grab what is left will create massive geopolitical changes. There will be more attempts to grab areas of the world where there may be oil.

In an attempt to delay the inevitable individual countries will do everything they can to grab more than their fair share of the remaining oil. America has already been doing this for most of a century.

In late June 2007 Russian President Vladimir Putin made an

extraordinary bid to claim a vast chunk of the Arctic as Russian territory.

A United Nations convention says that no one can claim jurisdiction over the Arctic area but Russian geologists claim that the Arctic is linked to the Siberian continental platform. Experts claim that the area contains ten billion tons of gas and oil and significant amounts of gold, nickel, platinum, lead, tin and other metals.

And, as it becomes clear that the oil is running out, there will be oil-grabbing wars. (I'll write more about this later in this book.)

But the land grabbing and the wars won't change anything. The best they will do is to delay the inevitable for the winners.

All this free energy has enabled us to increase our agricultural output and that, in turn, has enabled the world's population to grow to breaking point.

We have become addicted to the virtually free energy we've obtained from fossil fuels. All this free energy has enabled us to increase our agricultural output and that, in turn, has enabled the world's population to grow to breaking point. Today the whole world is addicted to and totally dependent on the energy it obtains from oil, gas, coal and other natural resources. We have become dependent upon fossil fuels in general, and oil in particular, and whether we withdraw from our dependence sanely and wisely, and in peace, is up to us.

The industrial society we know is coming to an end. Whether it comes to an end quietly and peacefully, or noisily and with violence, is, to a large extent up to us. But come to an end it will.

The chances, I'm afraid, are that the end will be increasingly violent.

The search for oil has dominated the last few decades.

The search for oil has led to war after war. Time and time again America has invaded strategically important countries, either in order to grab the oil or in order to grab land for pipelines. It was hardly surprising that the citizens of some countries should fight back, defend their culture and their lands against the Americans (and since its leaders chose to accompany America on its oil grabbing adventures, against Britain too).

Oil led to the American `war on terror' and subsequently gave birth to modern terrorism. Religious hatred was originally stirred up not by zealots in Iran but by oil hungry zealots in Washington. It is hardly surprising that we now live in a world of constant war and constant fear.

There is no little irony in the fact that oil, as well as giving us the opportunity to grow more food has also given us the weapons with which to fight with ever more ferocity.

Without oil there would be no tanks, no jet fighters and no bombers. Oil has given us the bus and the tractor. It has also given us the chance to spray napalm on Vietnamese children and to drop bombs on wedding parties in Afghanistan.

We tend to think that our society is indestructible. We forget that there have, in history, been many equally arrogant versions of civilisation, societies peopled by citizens who regarded their world as being similarly permanent and unassailable.

We forget (though given the events of recent years we really ought to remember) that nations come and go with remarkable ease. All civilisations are fragile and easily destroyed by unpredicted crises which no one predicted.

One popular modern theory is that big consequences are a result of big unforseen events such as invention of the wheel, the motor car, wireless, railways or the steam engine. Revolutions, usually unpredictable, can also have massive consequences, of course.

The discovery of fossil fuels in general and oil in particular was a big, unpredictable discovery that led to massive consequences.

But the running out of fossil fuels in general, and oil in particular, isn't unpredictable. On the contrary, it is entirely predictable.

And the consequences of the fuels running out will be just as massive as the consequences of their being discovered.

We forget what things were like before oil because no one alive can remember anything different to the way we live now. We like to think that we have a special place in history. We are the appointed ones. We like to believe that the other cultures which collapsed were primitive, pagan and technologically far inferior.

But why should our western society, our own current version of human life, be more indestructible than the mighty Roman Empire?

Science shows that civilisations are more fragile than we like to think.

In an article published in Scientific American, Michael Shermer studied the life span of sixty civilisations. He concluded that the average life span of a civilisation is just 421 years. He also discovered that modern civilisations don't last as the old ones used to last. It seems that, like everything else, civilisations just aren't built the way they used to be.

The average life span of the most recent 28 civilisations is only 305 years.

Modern civilisations have, of course, been more complex than their predecessors. Labour has been well developed. People have been trained and skills have been acquired. Modern civilisations tend to have various complicated levels of government and they usually have a hierarchical leadership structure. Most important of all modern civilisations tend to demand more natural resources to sustain themselves. Modern civilisations tend to be expensive to maintain. And so, when they run out of resources, they die.

Which brings us neatly to us.

We live in the energy civilisation.

And our world has been changed enormously by, and become utterly dependent upon, a single energy source: oil.

We are far more vulnerable than we think we are. Our society is, because it is the most complex, the most vulnerable society of all time.

And when our society collapses the result will be catastrophic. The end of our complex, interdependent society will result in massive financial hardship, in rioting, in revolution, in protests, in mass starvation.

In a book entitled `The Collapse of Complex Societies', American historian Joseph Tainter, suggested that modern societies collapse because they become so complex that they suffer from diminishing returns.

Jared Diamond, in a book called `Collapse', argues that most crises are caused by declining resources. He points out that whether or not a civilisation survives depends on whether or not its leaders recognise the problems and make the right decisions.

Time and time again history shows that leaders who are fixed in their views and who are unwilling to change will lead their societies further into disaster.

The people of Easter Island are a good example of a society which collapsed.

The leaders on Easter Island built up a tradition of erecting enormous statues. The bigger the statue the greater the status of the man who had it erected.

The problem was that erecting the statues used up huge amounts of timber, rope and human effort. Over three hundred years the Easter Islanders cut down all their trees. Every one. By the time they'd finished, their forested island had become unending pasture land. No more trees meant no more nuts. It meant the death of much wildlife. It meant that the Easter islanders could no longer build canoes with which to go fishing.

And it caused soil erosion which damaged crop yields.

The leaders of Easter Island were so stuck in their ways, so committed to the past, that they destroyed their own society.

They weren't the only ones.

Something similar happened to the Norse colony on Greenland. By cutting down too many trees to create pasture land for their animals the Greenland Norse found themselves without enough wood for building homes and for fuel. And without trees their topsoil eroded so their crops failed.

The Greenland Norse made other mistakes too.

Having come from Europe they insisted on maintaining their tradition of raising dairy cattle - even though that contributed to the deforestation. Worse still, the Greenland Norse failed to learn from their neighbours, the Inuit who had discovered their own ways of surviving in Greenland's cold climate. And, in an echo of the Easter Islanders folly, the Norse continued to spend huge amounts of their wealth on decorating their churches rather than on importing materials that might have helped them survive.

Throughout history there are similar examples. Even great civilisations such as the Romans failed because they refused to worry about the future and refused to change the way of life that had made them great.

And so to us.

Our global industrial system will probably collapse within the few decades.

Chapter Two: Peak Oil: The Beginning Of The End Of Civilisation

Vast quantities of crude oil come squirting out of the drill hole when oil is first discovered. That's the sight we're all accustomed to seeing in films. A black plume fired high into the sky. Gleeful oil prospectors dancing around, drenched in the black, sticky stuff.

To start with, getting the oil is easy. But suddenly the pressure starts to disappear and then the rest of the oil is harder to get out.

And then the oil starts to peter out. The point at which the oil starts to run out is the `peak' for that individual oil field.

The depletion of oil fields follows a predictable bell curve. The existence of this curve was first suggested by geophysics professor and Shell geologist, Dr Marion King Hubbert in 1956. It was Hubbert's way of predicting what would happen to petroleum production in America.

Hubbert's curve shows that production from any oil field rises rapidly, reaches a plateau and then falls quickly into decline. A graph of the rise and decline is called a `bell curve' because it looks like an upside down bell.

It was back in 1949 that geophysicist Marion King Hubbert first announced that the fossil fuel era would turn out to be very short. In 1956, based on a study of lifetime production profiles of oil reservoirs, and on reserve estimates, Hubbert announced that the peak of crude oil production in the USA would occur between 1966 and 1972. He predicted that oil production in the USA would peak in or around 1970. Everyone laughed. At the time the USA was the world's largest oil producer. And, oh how they laughed. Other geologists said it would be 1990 or even 2010 before America started to run out of oil.

American oil production didn't peak in 1970. (Although it wasn't clear until a year later that oil production in the USA had peaked.)

At that point, in 1971, oil wells throughout Texas started to dry up. American oil production turned down and never recovered. Gas and oil prices within the USA soared as American oil imports tripled.

Suddenly, and remarkably, everything had changed.

The Arab nations, through OPEC (Organisation of Petroleum Exporting Countries), suddenly had a hold over America. Geopolitics had changed for ever. Ever since then America has been reliant on oil imports and finding and securing the oil has dominated American foreign policy.

`It is evident that the fortunes of the world's human population, for better or for worse, are inextricably interrelated with the use that is made of energy resources,' wrote M.King Hubbert in 1969.

But Hubbert didn't just predict the end of American oil supplies. He also predicted similar peaks for other oil producing nations. And, one by one, he has been proved right. Venezuela hit its peak in 1970. Nigeria hit its peak in 1979. Many other countries have already peaked.

And, perhaps most importantly of all, the peak year for discovery was 1964. Despite spending countless millions, oil exploration companies have never again found as much oil as they did in 1964.

`We are in a crisis in the evolution of human society,' wrote Hubbert.`It's unique to both human and geologic history. It has never happened before and it can't possibly happen again. You can only use oil once. You can only use metals once. Soon all the oil is going to be burned and all the metals mined and scattered.'

Hubbert made another big prediction.

When he'd predicted what would happen in the USA and in other specific countries Hubbert looked at global oil production. He estimated that the peak would be at the end of the 20th century or early in the 21st century.

`...about 80% of the oil produced today flows from fields that were found before 1973, and the great majority of them are declining,' wrote Colin J.Campbell and Jean Laherrere, in Scientific American, in March 1998.`In the 1990s oil companies have discovered an average of seven billion barrels of oil; last year they drained three times that much. Yet official figures indicated that proved resources ...expanded by 11 billion barrels of oil. One reason is that several dozen governments opted not to report declines in their reserves, perhaps to enhance their political cachet and their ability to obtain loans. A more important cause of the expansion lies in revisions: oil companies replaced earlier estimates of the reserves left in many fields with higher numbers.'

The whole theory of peak oil is based upon Hubbert's law which effectively states that once half the oil is removed from an oil field the amount of oil which can be removed will decline, often quite rapidly.

Technology can help make sure that we get most of the remaining oil. But no amount of technology can increase the amount of oil available. And as the amount of oil (or gas or coal or whatever) remaining diminishes so it becomes harder to extract what remains. And, as it becomes harder to extract the remaining oil (or gas or coal or whatever) so it also become more expensive and it also uses more energy.

In order to get the best out of a field numerous wells have to be drilled. As more wells are drilled so more oil is obtained. Oil production peaks. Eventually, however, it gets harder to get at the remaining oil. Production falls off. Even if additional wells are drilled it is difficult to get the remaining oil out of the ground. And much of the oil will never be taken out.

During the 1970s (by no means the best decade for finding oil) 220 giant oil fields were discovered.

So far this century just 72 oil fields have been found. None of them have been large. The Centre for Global Energy Studies calculates that the average output from giant oil fiends found in the 1990s is about half that of the fields found in the 1970s.

According to the Oil and Gas Journal the average field size has been falling dramatically since the peak was reached in around 1940. Then the average field size contained nearly five billion barrels of oil. Today the average new field contains a fifth that much.

The result is that combined global oil and gas reserves have fallen dramatically from a peak in the 1960s and 1970s.

It is widely believed among people in the industry that OPEC has been deliberately inflating its estimates of its reserves to stop panic and to keep people using oil.

When will the party be over?

It is impossible to say precisely when the oil will run out and our civilisation will end. All it is possible to say with some certainty is that it will end - and much sooner than most people expect. According to the International Energy Agency (the industrialised countries' energy watchdog and not an organisation known for its pessimism as far as oil supplies are concerned) oil will start to be in real short supply by 2012. There will, says the agency, be a real `supply crunch'.

Once oil is in short supplies prices will rise dramatically. And they won't ever come down again. The coming disaster, the end of oil, will have begun.

Most of the oil which is exported to America and to Europe comes from countries with weak, developing economies and weak political structures.

Countries which are developed, or which are growing, need all the oil they can produce for themselves. So, although both the USA and the UK are oil producers, neither produce enough oil for their needs and so both countries are net importers.

Once an oil exporter becomes developed it invariably needs all the oil it can produce for its own industries and for its own transport.

So, for example, consider Iran.

Iran is one of the world's major sources of oil. It is the second largest exporter of oil in OPEC. And yet Iran claims that it needs nuclear reactors because it too will soon be an importer of oil.

If Iran needs to import oil just where will it get it from?

Writing in The Party's Over, Richard Heinberg argues that the end will come in several major steps.

1. The Peak in energy production

During the years from 1945 to 1973 world energy production per head of population increased by 3.24 % a year. From 1973 to 1979 the increase slowed to 0.64% a year. From 1979 energy production steadily declined. The conclusion is that 1979 was the peak year for energy production.

2. Peak availability of energy

For the last 20 years more energy has been produced each year. But the amount of energy consumed in obtaining that energy has increased more rapidly. It now takes far more drilling effort to obtain oil, natural gas or coal than it did a year or two ago. The peak years were, says Heinberg, probably between 1985 and 1995. (He points out that no official agency has bothered to make the necessary calculations.)

3. Peak global extraction (`peak oil')

Heinberg suggests that the peak year for taking oil out of the earth will probably be reached between 2006 and 2010.

Most politicians and so called experts continue to reassure us that soaring energy prices are a temporary problem, that oil reserves and effectively limitless and that we don't have to worry about oil running out in our lifetimes. Even if it does, they say, it won't matter. There are dozens of other new technologies available to replace the oil.

The politicians claim that when materials or commodities are in short supply people find ways to cope without them. And, they say, when prices rise adventurous companies will find ways to produce alternatives. If metals are running out, exploration companies will find more of them. If oil is in short supply and the price rockets, oil exploration companies will make sure they find some more of the stuff. They will do this because it will be in their financial interest to do so. High prices restrict demand and encourage supply.

People have, they say, been predicting the end of natural resources since the 19th century. And they are right about this. In 1865 a distinguished British economist called Stanley Jevons published a comprehensive study of the coal industry in which he predicted that Britain would need 2,500 million tons of coal a year by 1961. There wouldn't be enough of the stuff, he said.

As it happened our consumption of coal fell from a peak of 300 million tons in 1913. By 1961 it was down to 200 million tons a year. A tenth of the Jevons estimate.

What Jevons hadn't realised was that someone would find huge amounts of oil in the Arabian deserts, thereby making coal rather less important.

In 1972 the Club of Rome published a book predicting that the world would soon run out of gold, silver, zinc and other metals. Non renewable resources would run out and there would be worldwide chaos.

Wrong again.

Sometimes, even oil experts seem to agree with the optimistic politicians.

BP's Statistical Review of World Energy, published in June 2007, suggested that there is enough oil left for another 40 years of consumption at present rates. (Back in 1980 experts forecast that there was enough oil left for another 40 years of consumption. Is there always going to be another 40 years worth of oil?)

But the BP Review is a summary of estimates supplied by Governments and oil companies. These are often politically motivated.

It's worth remembering that Britain's North Sea oil reserves peaked in 1999 but the British Government wouldn't admit that this had happened for some years afterwards.

All oil producing governments lie about their oil reserves.

So, are the politicians and oil producing countries right now?

Or could they possibly be lying?

If the politicians are right then you have nothing to worry about. If you are prepared to trust their reassurances you can carry on with your life, content in the knowledge that the problems ahead of us will all be controlled if only we can learn to put our rubbish out on the right days and separate our plastic bottles from our glass ones.

On the other hand, if the independent oil experts are right and the politicians are wrong then the disaster heading our way is far bigger than anything any of us has ever envisaged.

The problem is that although the politicians are reassuring the geologists and oil experts are not. Nor, indeed, are the oil producers themselves.

By the turn of the millennium every key oil producer around the world had started to acknowledge that they were producing more or less at peak sustainable rates. Many of OPEC's key oil fields were in permanent decline and some OPEC countries had stopped exporting oil at all. In 2004, for example, Indonesia became an oil importer for the first time in its history. By the middle of 2004 there were concerns that the Russians were overproducing oil, and working their oil fields too hard, in order to maximise their current profits. (Russia has developed close links with India and China because whereas Russia has oil and no money, India and China have bucket loads of money but no oil.)

For twenty years the world has been finding less oil than it has been using. Demand has been soaring but the dribs and drabs of new oil that have been discovered have been found in out of the way, difficult to reach places. There have been no world-class oil discoveries since the finds in Alaska and the North Sea in the 1970s. Despite the huge amounts of money spent on drilling and exploration no major new oil fields have been discovered for twenty years. In not much over a century we have used up oil reserves which took millennia to form. Most of the oil we are burning was created over about 500 million years. Nothing can replace the oil we have taken out of the ground and used up.

Out of the 98 oil producing nations 60 are in decline.

Oil companies are spending more and more money in their search for oil. Between 2000 and 2005 they increased their capital expenditure by a quarter of a trillion dollars. We are, however, discovering less oil than at any time since 1945.

We have better technology than ever - including satellite mapping - but we are finding much less oil and we are, indeed, finding far less oil than we are using.

In the 1960s oil company geologists discovered around 55 billions of oil a year. Today they are lucky if they discover 9 billion barrels of oil a year.

The discovery of what oil experts now still regard as the `supergiant' fields began in the 1930s. The al-Burgan field in Kuwait (currently still the world's second largest oil field) was discovered in 1938 and has been in constant service ever since. The Saudi al-Ghawar field (still the world's largest oil field) has been slowly emptied since 1948. There are only 14 oil basins in the world which are capable of producing half a million barrels a day. All were discovered in the 1940s or earlier.

The days of sticking a well into the ground and then standing back as the oil gushes out are long gone. Today, the oil that is discovered tends to be the type of stuff that takes a good deal of (expensive) work to extract and refine.

Oil production by non OPEC nations reached its maximum in 1999 and has been in decline since then. Scientists at the Oil Depletion Analysis Centre say that global production of oil will peak by 2011 before entering on a steep decline.

They say that the peak of regular oil - the stuff that is easy to get out of the ground - went in 2005.

When you factor in the more difficult stuff - polar reserves and liquid taken from gas, peak oil will come by 2011.

Many individual oil producing countries have already reached `peak oil' production. They are now pumping out less oil than they used to. A number of these countries are, therefore, now net oil importers instead of being oil exporters.

Here are the dates when specific nations reached `peak oil'.

America   1970

Argentina 1998

Australia 2000

Congo 1999

Egypt 1993

Indonesia 1977

Iran 1974

Iraq 1979

Kuwait 1972

Libya 1970

Norway   2001

Oman 2001

Peru 1980

Saudi Arabia 1979

Soviet Union 1987

UK 1999

Venezuela 1970

Most other oil producing nations are expected to reach their individual `peak oil' year well before 2010.

Whenever there is a temporary shortage of oil, the prices skyrocket. This isn't something that only happens to oil, of course. It's a general rule of life. When there is a shortage of something the price goes up. When the stuff that is in short supply is something that a lot of people really, really want then the price will go through the roof. And there is panic and chaos. In 1970 a reduction of just 5% in the amount of oil available caused a price rise of 400%.

So, what will happen when the world's oil production starts to fall permanently?

Most, if not all, oil producers exaggerate their reserves.

In 1986 OPEC made a rule that its members could only export oil according to the amount they had in their reserves. Countries could only sell a proportion of what they had in the ground.

Guess what happened?

Within weeks of the new rule being introduced almost every OPEC country suddenly discovered that it had more oil than it had previously thought it had. Reserves were upgraded so that individual countries could sell more oil and earn more money. The upgrades came without any new oil wells being discovered.

The total reserves claimed by OPEC - the official amount of oil remaining - leapt from 353 billion barrels of oil to 643 billion barrels of oil.

The Saudis increased their estimate of their oil by 100 billion barrels. Kuwait added 50% to its reserves. Just like that. Whoomp. Venezuela doubled its reserves. Iraq and Iran doubled their reserves too.

No one had found significant new amounts of oil.

They just altered their official estimates of the oil they had left.

Countries are still exaggerating their reserves.

OPEC members have kept their reserve estimates the same year after year. They simply pretend that the oil they've sold has come out of thin air.

Every oil well in the world is apparently bottomless.

Speaking on Channel 4 news in 2004 Sadad Al Husseini, formerly vice president for exploration of the Saudi oil company Aramco, said that official American forecasts for future oil supplies are `a dangerous over-estimate'.

When he was asked if people should be worried Mr Al Husseini said they should.

Even companies exaggerate their reserves.

In 2004 one of the world's biggest companies, Royal Dutch Shell, admitted that it had overestimated its oil reserves by 4.5 billion barrels. During one remarkable short period Shell reduced its reported reserves three times, reducing the total amount of oil it said it had by a massive 20%.

Astonishingly, Shell executives had apparently assumed that they would be able to maintain and even increase extraction rates from their oil fields in Oman - despite the fact that production had been falling for four years. Their reserves were based not on oil they knew they had but on oil they expected to have. Similar revisions apparently had to be made for the oil the company had previously said it had in Nigeria.

It seems to me that it is as though you and I had included next week's expected lottery win in a statement of our resources when applying for a bank loan.

Worst of all is the fact that Shell was not, and is not, alone in overestimating its reserves. Other oil companies do the same thing.

`The ability to control energy, whether it be making wood fires of building power plants, is a prerequisite for civilisation,' wrote
Isaac Asimov in 1991.

Oil producing countries (and companies) are relying on new technologies for extracting oil to enable them to fulfil their published figures for oil production.

No one seems to realise (or, if they do, they don't seem to care) that taking oil out more efficiently doesn't increase the amount that of oil that is available.

If you take the oil of your oil fields more speedily then they will empty more speedily.

The four countries with the biggest reported reserves in the world are Saudi Arabia, Iran, Iraq and Kuwait.

Of these Saudi Arabia is regarded as the world's most important source of oil. The biggest oil field in Saudi Arabia is known as Ghawar. It is the largest old fashioned oil field in the world. (The oil sands in Canada are bigger but far less useful as a source of oil.)

In 1948 Ghawar was estimated to hold 87 billion barrels of oil. In 1970, experts from four top oil companies - Chevron, Exxon, Mobil and Texaco - estimated that there were 60 billions of oil left in Ghawar.

Since then 55 billion barrels of crude oil have been taken out of Ghawar.

Do the sums.

Ghawar now has no more than 5 billion barrels of crude left.

Every day the Saudis pump 7 millions of seawater into the Ghawar oil field to keep the pumping pressure high enough to get out the remaining dregs of oil.

Throughout the last half of the twentieth century the oil produced from Ghawar, made up between 55% and 65% of Saudi Arabia's total oil production.

For years the world has believed that there is so much oil underneath Saudi Arabia that it can never run out. Sadly, there has never been any reliable evidence to sustain this optimistic view.

`Twilight in the Desert' is surely one of the most important books to have been published in the last half a century. The author, banker and oil expert Matthew R.Simmons, explains the size of the threat facing the world in precise and unemotional detail. Matthew Simmons is an investment banker who manages over $50 billion in energy investments. He has been a White House adviser to both Bush and Clinton. Simmons has examined the fading of Saudi Arabia's oil supply and suggests that it is the coming failure of Saudi Arabian oil supplies which will bring about the peaking of global oil supplies - at the very time that the global demand for oil is rising faster than ever before.

Simmons points out that 90% of all the oil that Saudi Arabia has ever produced has come from just seven giant oil fields which have now all matured and grown old. The three most important oil fields in Saudi Arabia have been producing oil at very high rates for over half a century. Ghawar, the biggest single oil field, is known as the King of oil fields .

But now the end is nigh.

`Twilight at Ghawar is fast approaching,' says Simmons.

Just how quickly the oil will run out is something no one knows. Once peaking occurs, however, the decline in output tends to fall. It doesn't reach a plateau and just stay there. It is not uncommon for the output from a major oil field to halve in just a few years. Indeed, it is the norm. A survey of major oil fields showed that all of them declined by more than 50% within ten years of peaking.

The Brent oil field was producing 450,000 thousand barrels of oil a day in 1985 but, just five years later, in 1990, the production was down to 100,000 thousand barrels of oil a day. Both the Brent oil field and the Forties oil field (the UK's two largest oilfields and the mainstays of the North Sea oil field industry) were largely empty by the year 2000 - around 30 years after their discovery.

The Samotlor oilfield was producing nearly 3 million barrels of oil a day in 1986 but by 1994 the oil production was down to below half a million barrels of oil a day.

China's need to import oil stems not only from the fact that the nation's demand is rising rapidly. The huge Chinese Daqing oil field has been producing over one million barrels of oil a day for over 35 years but in early 2004 China's energy planners publicly discussed the likelihood that Daqing's output would be down by 40% by 2006 or 2007.

If Ghawar and other major oil fields fall only as quickly as the world's other oil fields (and, remember, Ghawar and others have been worked very hard for many decades and it is not unreasonable to expect that they could fall even more rapidly) the world oil supply will be devastated within less than a decade.

Remember that despite the spending of huge amounts of money on looking for oil there have been no major oil field discoveries for years.

The chances are that it will not be possible to replace dying oil fields with new ones. And so the oil age will end.

The oil produced in Saudi Arabia is crucial largely because of the quantities involved. For decades the oil produced by the Saudis has made up around half of the oil produced in the whole of the middle east. Saudi oil reserves make up nearly a quarter of the world's oil reserves. And, over the years, the willingness of the Saudis to increase their production whenever the world is short of oil has helped to even out the global supply. Today, the Saudis are paying the price for all those years of overproduction.

At a secret 1974 USA Senate hearing investigative it was reported that back in 1972 the American oil companies in Saudi Arabia had begun to realise that they were taking so much oil out of the Saudi Arabian oil fields that they were damaging the fields.

However, since the American oil companies knew that they were soon going to lose ownership of the oil fields to the Saudi Arabian

Government they deliberately decided to 'milk these fields for every saleable drop of oil and put back as little investment as possible'.

This allegation was repeated in 1979 when investigative reporter Seymour Hersh published an article in the New York Times questioning the capacity of Saudi Arabia's oil fields. Hersh claimed that the Saudi Arabian oil fields had been systemically over produced in the early years of the 1970s because representatives of America's major oil companies (which at that time controlled Saudi Arabia's oil fields) suspected that the Saudi Arabian Government was about to nationalise their oil fields. The American oil companies wanted to get oil out of the ground as rapidly as they could - without worrying about the fact that if you take too much oil out of an oil field you risk permanently damaging it.

In 1979 the US Senate Subcommittee on International Economic Policy of the Committee on Foreign Relations took another look at the Saudi Arabian oil fields. The Senate's advisers suggested that the 9.8 million barrel per day production rate was probably as good as it ever going to get. In other words, even if money was invested in them, the key oil fields in Saudia Arabia would all be in decline before the year 2000. This information was considered too alarming for the public and so the source documents for this conclusion were sealed from the public view for another 50 years - until long after the prediction would have been proved either right or wrong, and probably long after those who were responsible for it had gone on to worry about other things.

Predictions are only predictions, of course. Good guesses. Estimates. But this one seems to have been spot on.

By the end of the 1990s almost all OPEC producers were close to their peak production rates. Saudi Arabia had managed to increase its oil production in the 1990s in order to help the world cope with the embargo on Iraq's oil exports.

But it is likely that by doing this the Saudi Arabians damaged the long term viability of their oil fields. When you take extra oil out of an oil field you reduce the reservoir pressure and allow more unwanted water to seep into the remaining oil.

So, here's a piece of irony. The American sanctions, designed to force the Iraqis to let the Americans have access to the Iraqi oil, probably resulted in permanent damage being done to the huge Saudi oil fields.

The increasing fall in supply from most countries around the world has coincided with a sustained increase in demand and Saudia Arabia, which long ago gave itself the task of filling in when world oil supplies look insufficient, is now probably causing permanent damage to its oil fields by increasing production in order to keep up with the demand for oil.

When the valves are opened too wide (to get out extra oil) the pressure within an oil field falls. When the pressure falls it becomes increasingly difficult to get oil out in the future. More importantly, as the oil is drained out, water tends to pour into the void - further damaging the oil field.

The Saudi Arabians are not the only ones who are running out of oil. The world's second biggest oil field is the Burgan oil field in Kuwait. For almost sixty years this oil field has been pumping out vast quantities of oil. It accounts for over half of Kuwait's proven oil reserves.

But Kuwait recently revealed that the Burgan oil field is past its peak output. The oil field's planned output will drop from two million barrels a day to around 1.7 million barrels a day. And the fall will continue - probably quite steeply. In 2006 a journalist found documents suggesting that the country's real reserves were half of what had been reported.

In 2007 Iran became the first major oil producer to introduce oil rationing. And, of course, the Iranians want nuclear power to replace their diminishing oil supply.

And in Iraq the Americans have created chaos, constant war and instability.

The Americans expected to be able to take control of Iraq's remaining oil fields when they invaded in 2003 but things haven't worked out well for them. Resistance workers fighting against the occupation forces have repeatedly damaged pipelines and refineries and have been so successful that they have reduced the amount of oil being produced.

The factor which makes the crisis still worse than it might otherwise be is, of course, the rate at which the consumption of oil is rising.

Governments in Europe are desperately trying to persuade their citizens to use less energy. They're doing this not because the oil is running out (I haven't heard one Minister talk about peak oil) but because of global warming.

Unfortunately, the efforts in Europe aren't going to make much (if any) difference for two reasons.

First, the Americans aren't doing anything to cut back on their consumption of oil. According to the Energy Information Administration, America burns through 20 million barrels of oil a day. That's nearly a quarter of the oil used by the entire world.

Second, and probably even more important, parts of the world which never used to use much oil have discovered just how much joy the stuff can add to life. And they want some of the oil and some of the fun.

If China and India continue to grow (even at a much reduced rate) the world's consumption of oil will, inevitably, rise remorselessly. Even if America and Europe cut their consumption of oil (not particularly likely) the world consumption of oil will still rise. And it will keep rising.

At the moment the Chinese use, on average 1.5 barrels of oil per person per year. The average Indian uses less than one barrel per person a year. Contrast that with America where the average person uses 26 barrels of oil per person per year and the UK where the average person gets through just over 10 barrels of oil a year.

China and India need huge amounts of oil and their leaders are already doing everything they can to secure long term oil supplies. They are using diplomacy, trade, direct investment, technological assistance, military support, bilateral agreements - and all the other tricks previously used by Western countries. China has, for example, recently done deals with such vital oil producing countries as Iran, Saudi Arabia and Venezuela. Altogether, China has secured, or is in negotiations for, free-trade pacts with 25 countries. Two years ago China had no such trade agreements. China is positioning itself as a world leader in trade and investment and is beating the USA at all its own games. China has excellent relationships with North and South Korea, Pakistan and India and good relationships with the EU and with Australia and Canada.

Many countries now regard China as a better future ally than the USA and regard a relationship with China as a good defence against American hegemony.

The Chinese are in discussion with Canada about a joint effort to develop a number of energy-related projects, including developing the huge oil sands of Alberta.

Even if we had not reached `peak oil', even if the oil wasn't running out, even if the world's oil stocks somehow remained constant, the simple rules of supply and demand would ensure that the oil price would inevitably rise.

The only thing that would take the oil price downwards would be either discoveries of several massive new oil fields or the discovery of a genuine new alternative.

Since oil exploration companies have been searching for new oil for several decades without much success the former possibility seems unlikely. (I will deal with biofuels and the Canadian oil sands later.)

And to rely on someone somewhere discovering an entirely new form of energy is like putting all your money on a 1,000 to 1 outsider. It's possible but not very likely.

`According to the US Geological Survey, global discovery of large new oil fields peaked in 1962 and has been declining since,' wrote James J.MacKenzie, in Issues in Science and Technology dated June 22 1996.`The reason is simple: most oil occurs in a very few large oil fields and these are usually discovered early on because they are so big. The largest 1% of oil fields contain 75% of all the discovered oil, and the largest 3% contain 94% of the oil. The implication of this skewed distribution is that as exploration progresses, the average size of the fields discovered decreases. In other words, exploration in the declining phase of oil development - where we find ourselves today - is a far different game than in the early phase. In the early stages, it is the large fields that are readily discovered; in the declining stages, geologists are much more likely to find small fields and oil companies must do a lot more drilling just to stay even. That's why it's so much harder to maintain production in the declining stages than in the growing phase of the industry.'

The consensus among experts seems to be that if we haven't already reached the all time peak in global oil production (global `peak oil') then we will do so within the next year or two. And once we are over the peak the downward slide will be fast and constantly getting faster.

Current oil consumption levels around the world (measured in barrels per head per annum) are:

a) India 0.6

b) China 1.5

c) Russia 9.6

d) UK 10.4

e) USA 25.8

It's crucial to remember that the demand for oil is increasing in China at 30-40% a year. An increasing number of countries are now competing for shrinking supplies.

The oil sands of Alberta, Canada are said by those who sneer at the concept of peak oil to be the answer to the world's coming oil shortage. Despite the enormous problems the Canadian oil sands now account for more than 30% of Canada's oil production.

The Athabasca Oil Sands, for example, cover 141,000 square miles (which means they are cover a larger area than the whole of England). There are huge variations in the estimates of the oil available there but it is widely agreed that there is more oil in the Canadian oil sands than there is in Iraq. Theoretically, there is enough oil in Canada's oil sands to provide up to half of America's oil use. But the Americans have so annoyed the Canadians that it seems very likely that any surplus oil will go not to the USA but to China. (Serious rumours abound in North America that there are plans for America, Mexico and Canada to unite and create a single new country with a common energy policy. The aim, of course, is for America to use the merger to steal Canada's and Mexico's oil and gas resources. Both countries are, however, expected to reach their peak oil production in 2007 so if and when the Americans realise this the marriage may be off.)

Getting the oil out of these sands is difficult. The oil sands may contain the biggest known reserves of under-exploited oil in the world but the oil is trapped in a mixture of sand, water and clay. It is extremely difficult and expensive to get at the oil. And it is enormously damaging to the environment to do so. When the oil is removed it's a thick, heavy tar-like substance called bitumen.

The oil sands simply aren't as wonderful as some people think they are. Canada is sitting on a potential 315 billion barrels of oil but there are only two ways to get the oil out: either by mining it or by heating it deep underground so that it is forced up to the surface. The viscous tar that is produced then has to be processed before it can be refined.

Getting the oil out is, in short, expensive, messy and very, very bad for the environment. Getting the oil out also uses up a great deal of energy.

The sand has to be effectively `cooked' to get the oil out and it has been estimated that it takes the equivalent of two out of every three barrels of oil obtained from the sands to pay for the energy and other costs involved in getting the oil from the sands in the first place. Two tons of sands have to be mined to produce just one barrel of oil. Using natural gas to `cook' the sands is one possibility but there doesn't seem much point in using vast quantities of natural gas to release oil. Canada has some of the world's biggest deposits of uranium and it seems likely that Canada will use some of its uranium to get the oil out. Early in 2007 Canada's Natural Resources Minister agreed that Canada would probably use nuclear power to help get the oil out of the oil sands.

There are absolutely devastating environmental consequences. Huge, permanent lakes of toxic waste are created during this process. The world will undoubtedly continue to try to extract oil from the sands but the price will be a heavy one and the reward a rather light one.

Incidentally, Canada can't extra oil from the sand and stick to the Kyoto Protocol which it has signed. Canada is committed to reducing its greenhouse-gas emissions to 6% below its 1990 level of 560 million metric tons by 2012 but things are not looking too good. In 2003 Canada's emissions rose to 740 million metric tons. And the oil-from-the-sand industry is going to add around 82 million metric tons a year to that total.

`How do we stick to 1990 emissions levels when our population is greater and our opportunity is many times greater?' asks the Alberta energy minister Greg Melchin, who has obviously learned a thing or two about global responsibility from his neighbours to the south. Oil sands producers are already the largest and fastest rising source of greenhouse gases in Canada if not the world. By the year 2015 the oil sands will be emitting 156 million tonnes of greenhouse gases a year.

`All this energy expended in thousands of ways used to finally discover oil and produce it has to be added up and compared with the amount of energy in the oil which these efforts produce,' wrote Walter Youngquist in `Geodestinies: The Inevitable Control of Earth Resources over Nations and Individuals' in 1997.

So, is ethanol from corn the answer to our disappearing oil supplies?

It's been known for many years that you can run an engine on oil made from all sorts of things other than the black sticky stuff that comes out of the desert. At the 1898 Paris Exhibition an inventor demonstrated an engine which ran on 100% peanut oil. Vegetable oils were abandoned not because they don't work but because they are more expensive than the oil that comes out of the ground.

Now, with the oil running out, some people claim that we can solve global warming, and avoid the problems caused by peak oil, by burning ethanol made from corn (and other crops).

Ethanol has suddenly (and bizarrely) become everyone's favourite short cut to reducing emissions of greenhouse gases.   To   the Americans it is an easy way for them to remain addicted to oil without being hooked on imported oil.

The drive to replace oil with biofuels is strongest in America where George Bush has discovered new friends among the farmers who grow stuff which can be turned into biofuel.

Disguising this greed as a `cleaner energy initiative' America is aiming to increase the use of what they describe as 'renewable fuels' fivefold by 2020.

George W. Bush and other American politicians love biofuel because it justifies spending billions on farm subsidies (and, thereby, pleases farm backers and lobbyists who will take over from the oilmen).

Drivers love biofuels because, thanks to subsidies, they are cheaper at the pump.

The biofuel of the moment is ethanol - despite the fact that it is fuel and land intensive to produce. (It actually costs more to produce than petrol (about twice as much).)

Ethanol is fashionable because those who promote it claim that its use will save the planet from global warming.

Car makers in Detroit have agreed that half of their vehicles will run on a petrol-ethanol mix in the ratio of 85:15 by the year 2012.

The type of ethanol they prefer is the one made from corn. Naturally, this is the most costly and inefficient way to produce a biofuel. It is also the way that does most harm to the environment. Brazilian ethanol produced from Brazilian sugar cane is far more efficient but corn is grown in America by large corporations with lots of expensive lobbyists and Brazilian sugar cane is grown in Brazil.

And, as everyone knows, some powerful Americans don't give a fig about the environment. If it's a choice between making a buck and screwing the environment the Americans will screw the environment every time. The result is that a fifth of the American corn crop is already used to make ethanol.

There are 76 ethanol refineries under construction in the USA. In 2006 20% of America's corn crop went towards making ethanol and yet ethanol is still only 4.3% of the volume of petrol sold. So to replace gasoline will need 500% of the corn crop.

The corn producing states love ethanol and their lobbyists are replacing the oil company lobbyists in Washington. The American Government is happily throwing money at ethanol production because it makes them look as if they care about the fate of the planet. The American Government encourages the use of ethanol by giving a tax subsidy of 51 cents a gallon. Bush gave $2.5 billion to the oil companies to mix ethanol in with their petrol (which they would have happily done anyway). The Bush target is that by 2017 the entire current American crop of corn will be turned into fuel.

In fact the Americans would have to turn 100% of their soybean and corn crops into fuel to replace 11% of on road fuel consumption in America.

Just what people will eat is something the Americans haven't thought about yet.

The EU is as barmy as America.

The European Commission, father and mother of the straight banana and the perfect cucumber, and a safe but wildly overpaid working environment for many of the world's most blindingly stupid people, has called for biofuels to replace 10% of all petrol by the year 2020.

Can the European Commission have really matched George W.Bush for stupidity?

Oh yes.

To reach its target 25% of all European arable land will be turned over to ethanol production.

The consequences of this madness are seemingly unending.

Farmers who used to grow soybeans are now growing corn because they can either sell it as food or sell it to the oil companies to mix in with their petrol. Rapeseed crops are now seen everywhere in Europe as farmers abandon growing barley and wheat. (The disappearance of barley is, of course, a reason why beer prices are rising. And will continue to rise.)

I wonder how many farmers know (or care) that rapeseed oil produces lots of nitrous oxide gas - which creates even more global warming than carbon dioxide?

Those who believe that ethanol offers answers to all our problems are magnificently, technicolour stupid. Those who believe that ethanol offers any answers to some of our problems are just plain stupid.

The lunatic belief in ethanol has increased the price of food around the world and is boosting the price of farmland. The makers of Italian pasta, Mexican tortillas, American corned beef and German beer have all warned that the price of their products will have to rocket. Food conglomerates have issued a series of profit warnings. The ethanol promoters don't seem to care but their bright new idea will result in a massive increase in worldwide starvation.

The pro-ethanol strategy (as they like to think of it) is eye wateringly wasteful. It's so stupid that it's difficult to say how stupid it is. But I will try.

It's like trying to replace the world's shrinking glaciers by making ice cubes in a series of refrigerators, carting the ice cubes by lorry to the site of the shrunken glaciers, and then hoping that the ice cubes will repair the damage.

Here are several good reasons why biofuels won't work.

First, Creating ethanol from corn will use up vast amounts of our corn supply. The corn needed for one tankful of ethanol could feed one person for a year. In Iowa in the USA, 25 ethanol plants are now operating, and 30 being built or planned. Once built these plants will consume half the state's crop of corn. The prices for sugar, corn and wheat are rising because we are using these foodstuffs for fuel. A huge chunk of world grain consumption is now going into American petrol tanks.

(At a time, it should be noted, when America has already become a net importer of food.) This will result in greater starvation in a world where there is a continuing and (because of global warming) developing food shortage. The global lack of fresh water (caused by the increase in the number of bathrooms being built, the increase in the number of people on the planet, the increasing demands of providing water for increasing numbers of cattle, the increasing pollution of our water supplies and global warming) will make growing crops even more difficult. And will also result in higher prices for basic food stuffs.

Using rape-seed and sugar beet instead of corn doesn't help because (and this will doubtless come as something of a surprise to politicians and bureaucrats) you still need land to grow rape-seed and sugar beet.

And when you've grown the stuff you still need to use lots of energy to harvest it, transport it and turn it into ethanol. (The dullards at the European Commission probably imagine that you plant some sugar beet seeds and wait for a nice neat row of petrol pumps to pop up a few weeks later.)

Second, it costs more in terms of energy to get the fuel than it delivers. There are energy costs in building and running tractors, and moving the corn around. You have to include the energy needed to plant the corn, water it, harvest and turn into alcohol. Oil is needed to make the fertilisers and pesticides the farmers use.

Numerous experts have shown that producing ethanol actually costs energy.

According to David Pimentel, a professor of ecology and agricultural science at Cornell University in the USA, 129,600 British thermal units of energy are used to produce 1 gallon of ethanol. And one gallon of ethanol provides just 76,000 Btu of energy.

Another expert has worked out that around 131,000 BTUs are needed to make a gallon of ethanol but that gallon will only produce 77,000 BTU.

These figures means a loss of over 50,000 BTU for every gallon of ethanol produced.

Oil, in contrast, is just dug out of the ground. The most important thing to remember about fuels is the energy output to energy input ratio. This measures the amount of work you have to do in order to obtain the fuel. The energy output to input ratio for oil is magnificent. It varies between 30 to 1 and 200 to 1 depending on where the oil is. If it's easy to get at then the ratio is probably closer to 200 to 1. Ethanol is very different. It depends on plants which have to be grown afresh every year. To grow plants to turn into ethanol you have to plant seeds, harvest the crop, take the crop to the refinery and then start work on it. You need a tractor to plant the seed, a tractor to harvest the crop and a lorry to take the crop to the refinery. It takes fuel to create the fuel. Oh, and then there's the fertiliser, the herbicide and the pesticide the farmer uses. Where do those come from? Oil. I repeat: to obtain ethanol we have to use up energy to make energy. We get less out than we put in.

Third, burning plant based ethanol still contributes to global warming.

Fourth, pollution from ethanol could create worse health hazards than gasoline - especially for sufferers from respiratory diseases such as asthma. Ethanol burning cars increase the level of toxic ozone gas in environment. This will mean that the atmosphere will become much more dangerous. Pollution from ethanol is more dangerous than petrol because when it breaks down in atmosphere it produces considerably far more ozone than petrol does. Ozone is corrosive and damages the lungs. (Ozone is so corrosive it can crack rubber and destroy stone statues).

Other substances released from use of ethanol as fuel include benzene, butadene, formaldehyde and acetaldehyde. All these are carcinogens.

Fifth, the search for so called green fuels, with which we can all prevent climate change and save the environment, is resulting in some strangely destructive behaviour. So, for example, the rising demand for palm oil (an ingredient in biodiesel) has led to tropical forests being cleared in vast areas of South East Asia. People are now chopping down hardwood forests in Asia in order to grow palm oil to take to the USA for the biodiesel industry. This is being done in order to keep petrol prices down so that Americans don't have to change their way of life.

Encouragement (and subsidies) from the USA and the European Union means that farmers all around the world are now creating more farmland on which to grow sugar beet and rape seed and corn. Higher prices for their produce means that there is an incentive to clear land and plant crops.

That surely cannot be a bad thing, say the empty skulled bureaucrats in Brussels.

Oh dear.

It is.

Because in order to obtain fresh farmland the farmers of Brazil and Asia are chopping down vast tracts of rainforest.This, surprisingly enough, means less rain forest.

And, in turn, means that there are fewer trees to get rid of all the carbon dioxide being produced by the tractors and lorries used by the farmers planting the corn, the sugar beet and the rape-seed.

Clearing rainforests to increase the land available for the cultivation of palm oil is bringing ecological disaster for countries such as Indonesia and Malaysia. And it is adding to the global warming problem. Indonesia, a key palm oil producer, now has the worst carbon emissions level per head of population thanks to the fact that forests have been cut down to make room for palm oil production.

The United Nations has predicted that the natural rain forest of Indonesia will have disappeared in just 15 years time, because of the planting of palm oil to turn into biodiesel so that Europeans (and Americans) can keep on driving their motor cars and taking long flights. Plantations of oil palms for biodiesel have been held responsible for 87% of deforestation in Malaysia. The oh so bloody goody goody fuel planters are ploughing up the planet. As I write, their bulldozers are busy in Africa and South America as well as Asia. The environmental damage seems endless. Sugar cane being grown in huge amounts degrades soils, causes pollution when fields are burnt to get rid of stubble and destroys wildlife.

The United Nations has also reported that biofuels use up vast amounts of water (a commodity that is becoming increasingly scarce). And monocropping (growing the same crop year after year) increases pests and has a negative impact on soil quality.

The bottom line is that bioethanol and biodiesel are a huge scam which will increase not reduce greenhouse gas emissions. Indeed, biofuels may well be the final nail in the planet's coffin.

It's about time the self-righteous nonsense preached about ethanol was put to rest. There is nothing remotely green about biofuels such as ethanol. It is a grotesque myth that biofuels are carbon neutral and will help with the oil shortage.

Biofuels are promoted as planet friendly. But ethanol (from corn or sugar cane) and biodiesel (made from soybean or palm oil) are no answer to any of our problems. The only people benefiting are the motorists who are able to keep driving their cars, the governments who are able to keep collecting the taxes raised on those fuels and the corporations growing and selling the crops which are being used to make biofuels.

The only good thing to come out of all this is that OPEC has threatened to cut oil production if the West continues to use more ethanol. This will increase oil prices but it may help ensure that the oil lasts a year or two longer.

Anyone who thinks that biofuels are an answer to our problems is certifiable. The American Government and the EU are spending billions of taxpayers' dollars and euros subsidising the biofuel industry. And yet the OECD has calculated that it would take 70% of all Europe's farmland to supply enough biofuels to save 10% of the oil currently used in transport.

Will anyone take any notice of all these truths? Will the Americans stop their mad race to replace oil with biofuel?

Of course not.

That's just another reason why I believe the world is heading for the Oil Apocalypse.

`This much is certain,' wrote Kenneth S.Deffeys in `Hubbert's Peak: The Impending World Oil Shortage' in 2001, `no initiative put in place starting today can have a substantial effect on the peak production year. No Caspian Sea exploration, no drilling in the South China Sea, no SUV replacements, no renewable energy projects can be brought on a sufficient rate to avoid a bidding war for the remaining oil. At least, let's hope that the war is waged with cash instead of with nuclear warheads.'

The all time high for world oil production was 85 million barrels a day and it was reached in December 2005. Demand has exceeded that almost ever since but production has never again risen to that level.

America now imports 70% of the oil it consumes and the American Government, aware of what is happening, has stored around a four year supply of conventional crude oil in huge holes in the ground. Other countries - such as China - are now building up huge stockpiles of their own. This stockpiling will, inevitably, accelerate the rate at which the shortfall increases.

In 2004 Matthew Simmons said: `Oil is far too cheap at the moment...the figure I would use is around $182 a barrel.'

Since 2004 things have got steadily worse.

Ghawar isn't the only big field that is running dry. Many of the world's biggest oil fields are between 30 and 100 years old.

In the end we won't know that the world has reached `peak oil' until after it has happened. Oil prices will rise rather rapidly, and they won't go back. A study of production figures will show a decline. Any economic recession which produces a lowered energy demand will mask the fact that the oil is disappearing.

Every year after `peak oil' it will be impossible to pump out as much oil as was taken out of the ground the previous year. Oil will become scarcer year by year. And because it is becoming scarcer it will, inevitably, also become more expensive.

Energy obtained from oil and other fossil fuels is, at the moment, essential for just about every human endeavour. Cars, street lights, computers, lorries, aeroplanes, food, toys - everything we do, everything we make is reliant on fossil fuels.

So, what do we give up first?

How much more will you pay for your petrol so that you can continue driving? How much will you be prepared to pay to heat your home? How much more will you pay for your food? How will your employer cope with rising costs?

For the first few years after the decline starts there will be only be a little less oil (though increasing demand from China and India will exacerbate this natural phenomenon). But after a few years the fall off will become much more dramatic.

We have for over a century been accustomed to living in a world where the amount of free (or `cheap') energy increased every year. No more. Those days are over for ever.

In our modern, complex world, energy means oil. Without oil we can't drive our cars. Delivery men cannot drive their trucks and lorries. Farmers can't operate their tractors or their combine harvesters. Without oil the factories won't run and there will be a desperate shortage of electricity. Without electricity the lights will go out and the computers won't work.

It is, therefore, hardly surprising that for decades the oil price has been the most significant factor in deciding how well the economy does.

Generally speaking, when the oil price goes up the economy does badly.

And when the oil price goes down the economy does well.

Simple.

For example, in 1973 when, in retaliation for American support of Israel, the OPEC nations imposed restrictions on oil exports, there was an immediate 70% rise in global oil prices. Shortly after this, in a protest about America's support for Israel in the Yom Kippur War, OPEC pushed oil prices even higher by imposing a total embargo on oil exports. The price of oil went from $3 a barrel to $12 a barrel. (The Israelis had received massive military aid from the USA - including aircraft to replace the ones which the Egyptians had shot down - and the Arabs, not unreasonably, believed this had affected the war's outcome.) Oil prices rose fourfold and the result was economic chaos. Growth stopped. The stock market collapsed. In 1974 the world went into economic shock. It was the biggest economic shock since the 1930s.

American President Jimmy Carter subsequently made a speech proposing a campaign to reduce America's dependency on foreign oil. He promised that from that time forward America would never increase its foreign imports and that plans would be made to reduce energy consumption and to develop alternative fuels.

Naturally, since this was a politician's promise, none of this happened. Oil consumption doubled between 1985 and 2000 and the use of renewable energy hardly changed. Cars were made which were less fuel efficient than they were in the 1980s. And since 1994 the USA has been a major net importer of oil.

In 1979 things got even worse for the West. The Iranian people overthrew the Shah of Iran (who had been put on the throne by the Central Intelligence Agency in 1953) and then fought a war with Iraq. (The Americans encouraged the overthrow of the Shah because he'd wanted to negotiate a better deal with the oil companies. The Americans also encouraged the subsequent Iran-Iraq war because they wanted to see both countries weakened.) The Americans hadn't realised the effect this would have on the oil price which went from $12 a barrel to $30 a barrel.

Since the rise in the oil price had been produced by artificial and temporary problems the oil price went back down again in the 1980s.

There is a widely held theory (popular among politicians and bankers) that oil price will not, in future, behave like other substances (which go up when they are in short supply and go down when there is a glut) but will merely oscillate around a mean point.

The theory says that commodities like oil have a `normalised' price which will always stay within a trading range.   The general view is that although the oil price might occasionally rise up as high as $105 a barrel it will soon come back down again to its trading range of around $40 to $50 a barrel.

This theory is put forward even by experts who acknowledge that the world demand for oil is rocketing (as China and India and other countries become more westernised in their approach to the use of oil burning products such as motor cars and aeroplanes) at the same time as the oil producers are finding it impossible to increase their production levels.

I find this belief touching but rather naive. I suspect that those who subscribe to it are also devotees of the tooth fairy and believers in Father Christmas.

As citizens and taxpayers we pay our politicians to look ahead, warn us about impending crises and, where possible, to find solutions.

But our politicians won't even admit that there might be a problem.

Instead of showing any concern for our long term security and prosperity governments everywhere prefer to ignore this problem. At the very best we get patronising reassurances.

The Department of Energy in the USA has responded to fears about the future supply of oil by predicting that the USA will be able to produce more oil in the future. They have claimed that American oil production will rise by 7% over the next few years. This is a heroic assumption, given that American oil production has been in decline for the last 35 years and there are no signs of any impressive new oil fields on the horizon. America's Department of energy has also stated (without any evidence for the claim) that the Saudis will be able to raise their production of oil in the future.

It's difficult to avoid the conclusion that wishful thinking is being preferred to scientific assessment.

The reality is that the oil price is never going back below $50 a barrel for any sustained period.

On the contrary it will continue to edge upwards. First, it will break through the $100 a barrel barrier. Then the $200 barrel barrier will be hit.

Even when the price hits $260 a barrel the price will still only be equivalent to the highs of the early 1980s.

Between 1970 and 1982 the price of oil went from $1.35 a barrel to nearly £35 a barrel. That's a twenty six fold increase. If the oil price went up twenty six times today (at the time of writing) it would be $2,000 a barrel.

During the early part of 2007 the price of oil was kept low artificially.

The Saudis wanted oil prices kept down to discourage the development of alternative sources of energy and to avoid conservation of supplies. The Saudi ruling family wanted to turn their oil into money as quickly as possible.

When oil is really in short supply the price will rocket.

What will the world's billionaires be prepared to pay for the oil to fuel their private aeroplanes and their Ferraris?

How about £20 a gallon? £50 a gallon? £100 a gallon?

And, remember, when the price goes up it will not be a spike or a blip.

The oil shortage will be permanent.

And so the oil price will never come down again.

The oil isn't going to run out overnight.

But once it becomes clear that oil is in permanent short supply, several things will happen to push up the price.

First, governments all around the world will start accumulating and storing oil.

Second, the military in every country will demand its own stockpiles. No country is going to want to have aeroplanes which can't take off because there is no fuel.

Third, speculators will start buying oil.

At that point a chart showing the price of oil will consist of a line going virtually straight upwards.

At what point will petrol become too expensive for the average motorist? When it costs £200 to fill the petrol tank? £500? £1000? And at what point will the central heating be turned off? When the bill reaches £200 a week? Or only when it gets really expensive?

Under normal circumstances governments use various methods to control the economy.

Interest rates can be lowered. This tends to encourage people to borrow money and, therefore, to spend it. This is usually done if the economy is stagnating or faltering.

Interest rates can be raised. This tends to discourage borrowing, slow down spending and take the heat out of an overheating economy.

The government can increase its own spending. It can build more roads, more hospitals and more schools. It can hire more employees. It can start a war. (Wars stimulate the economy by putting more money into arms production. And, of course, when weapons are fired the bullets and rockets have to be constantly replaced.) All these things tend to stimulate the economy by putting more money into the hands of the people.

The government can reduce its own spending. It can cut back on spending programmes and stop hiring new employees. This tends to put a damper on the economy.

The government can reduce taxes. This means that people have more money in their own pockets. More money means more spending. And that boosts the economy.

The government can increase taxes. Making people hand over more of their earnings means that there is less to spend and so the economy will slow down.

Those are the theories.

In practice a lot of things can go wrong. And there are a lot of possible complications.

So, for example, one problem is that if a country has little or no manufacturing industry of its own then allowing people to spend more money will mean more imports. And that will have an effect on the balance of payments. If a country spends more on imports than it makes by exporting then it will have a balance of payments problem.

Another problem is that if a country puts up interest rates this encourages foreigners to invest their money in the banks of that country. If Britain has higher interest rates than, say, Australia, then Australians who want a higher rate of interest on their savings will put their money into sterling. This will strengthen the British currency. For holiday makers going abroad that it's a good thing. It means that they will get more euros or dollars for their pounds. But for exporters, trying to sell in foreign markets, its a bad thing. It means that British made items become more expensive.

Rising energy costs are a particular problem because they do two things at once.

The first thing they do is add to inflation.

When the oil price goes up the cost of a gallon of petrol goes up too. Each time the price of oil goes up it has a bigger and bigger effect on the price of petrol or heating oil. This is because the other costs (such as the retailer's profit, advertising and taxation) will probably remain fairly stable. And so as the price of oil goes up and up so the cost of a gallon of petrol (or heating oil) increases much more rapidly.

Eventually, when the price of oil gets high enough all the other costs (retailer's profit, advertising and taxation) will be so small as to be irrelevant. (In practice, of course, these may change. Governments may attempt to take advantage of the rising oil price by hiding a tax increase in the price.)

The rise in the price of petrol and heating oil aren't the only things that will happen, of course. Oil is used for making plastic, fertilizer, asphalt and a million other things. It's also used to operate delivery lorries. Ships use it. So do aeroplanes. And tractors need it too.

And so many other prices will rise. And will add to the real rate of inflation.

In the past resource shortages have invariably led to double digit levels of inflation. So, as the oil price goes up we can expect to see inflation going up to 15%, 20% or even higher.

Normally, when there is inflation there is also growth. Prices rise because everything in the economy is going well.

But when inflation is going up because of rising oil prices workers will want more money in order to be able to buy the same things that they could buy before. And rising wages will discourage employers from hiring because they are already having a hard time coping with the higher fuel and material costs. And so unemployment will rise.

A temporary rise in the oil price can do an enormous amount of damage to the economy. In the 1970s rising energy costs (caused by OPEC cutting back production) resulted in inflation, unemployment and recession. The word `stagflation' was invented to describe this strange mixture of inflation and deflation.

A permanent rise in the oil price can do much more harm than a temporary spike. A permanent rise in the oil price will destroy an economy - permanently.

Governments will be faced with the choice of either trying to fight inflation by pushing up interest rates and cutting Government spending or of stimulating growth in order to try and boost employment.

If a government chooses the first of these the problems will be exacerbated by any national debt. In Britain, for example, where huge debts were built up during the late 1990s and the early part of the 21st century, the result will be that if interest rates are pushed up (say to 15% - a level they reached just a couple of decades ago) then house prices will go through the floor and millions of people will see their homes repossessed. There will, quite literally, be millions of people living on the streets. As house prices collapse so the stockmarket will follow suit. And unemployment levels will rocket. The Government will be unable to raise enough money to pay its wage and pension bills and millions of Government employees will find themselves joining private sector workers and pensioners on the streets.

If a Government chooses instead to stimulate growth they will send inflation soaring to unprecedented levels. The million pound postage stamp will be a real possibility. And I'm talking about the stamp you put on a letter, not the stamp you buy at auction. If this happens then savings will be worthless.

Some people will benefit (in the short term, at least).

Home owners will see the value of their properties rocketing. And governments will see their own debts disappearing. A debt of billions will be of little consequence when it costs £10,000,000 to buy a newspaper or a loaf of bread.

The consequences of this sort of inflation can be devastating. It was, remember, the inflation that destroyed the middle classes in Germany in the 1920s which gave Hitler the chance to take power.

(There are rumours that the Americans are already planning to dump the American dollar and create a new currency when all this happens to them. It is perfectly possible that the British Government would take the same option - either revaluing the currency or simply getting rid of the old `pound' and creating a new one.)

In the end nothing a government does will make any difference, of course, because when there isn't any more oil there won't be any more oil.

Each economy will be an unmendable mess and our civilisations, the ones we know and are accustomed to, will not survive.

Chapter Three: Oil Wars: Past, Present and Future

Peak oil has already been directly responsible for our loss of freedom.

Securing oil supplies was an important element in many of the wars of the twentieth century. It was certainly the major factor in America's recent illegal wars. The war against terrorism was merely a convenient and publicly acceptable excuse for unacceptable behaviour.

`The life contest is primarily a competition for available energy,' wrote Ludwig Boltzman in 1886.

Fighting for oil isn't new, of course.

America only entered World War I (on the side of Britain and France) after both its new allies and new enemies were pretty much exhausted by the fighting. Once it agreed to join in the war America imposed conditions which included the demand that America's economic and political objectives be taken into account when the war was over. One of those objectives was access to new sources of raw materials, particularly oil. In February 1919, Sir Arthur Hirtzel, a leading British official warned: `It should be borne in mind that the Standard Oil Company is very anxious to take over Iraq.'

That was 1919.

America demanded that its oil companies be allowed to negotiate freely with the new puppet monarchy of King Faisal (the monarch whom the British had put on the throne in Iraq). And so Iraq's oil was divided up between the allies. Five per cent of the oil went to an oil magnate called Gulbenkian (known as `Mr Five Per Cent') who had helped negotiate the agreement. The other 95% was split four ways between Britain, France, Holland and the United States of America. Companies now known as British Petroleum, Shell, Mobil and Exxon pretty much had a monopoly of the oil available. Iraqi oil was split this way until 1958 when there was a revolution in Iraq.

`Oil has literally made (American) foreign and security policy for decades,' said Bill Richardson, American Secretary of Energy in 1999. `Just since the turn of this century, it has provoked the division of the Middle East after world War I; aroused Germany and Japan to extend their tentacles beyond their borders; the Arab Oil Embargo; Iran versus Iraq; the Gulf War. This is all clear.'

American influence in the region was sealed when the al-Saud family and the United States of America created Saudi Arabia in the 1930s, pretty much as an American colony. It was no coincidence that the American Embassy in Riyadh, the capital city, was situated in the local oil company building.

The Americans were not, however, satisfied with their share of Middle East oil. They wanted control. They had to get rid of the British. And their chance came with the Second World War.

The Americans unceasingly portray themselves as Britain's saviour. This is a wicked misrepresentation. As it had been in the Great War, America was ruthlessly opportunistic.

Britain was greatly weakened by the Second World War but America grew tremendously in power as a result of what happened in the early 1940s. The Roosevelt and Truman administrations (which were dominated by banking and oil interests) decided to restructure the world to ensure that the USA would be on top. They wanted control of the world's oil. They wanted USA dominated globalisation (to which end they created the International Monetary Fund and the World Bank in 1944). They wanted the dollar to be the only significant world currency. And they wanted the USA to have military superiority in all types of weapons.

Winston Churchill was so worried by what he could see happening that on March 4th 1944 (three months before the D Day invasion of Normandy) he sought assurance from the USA that she would not try to take over British oil interests.

He wrote to USA president Roosevelt saying: `Thank you very much for your assurances about no sheep's eyes on our oilfields in Iran and Iraq. Let me reciprocate by giving you the fullest assurance that we have no thought of trying to horn in upon your interests or property in Saudi Arabia. My position in this, as in all matters, is that Great Britain seeks no advantage, territorial or otherwise, as a result of this war. On the other hand, she will not be deprived of anything which rightly belongs to her after having given her best services to the good cause, at least not so long as your humble servant is entrusted with the conduct of affairs.'

Sadly, there was nothing that even Churchill could do to save Britain from its new `enemy'.

The Americans had already acquired a new `special relationship' with Saudi Arabia. They arranged this in 1945. Since then the Saudis have helped the Americans by controlling world oil prices to the advantage of the Americans (by releasing or withholding oil supplies) and by continuing to sell oil in dollars (when other oil producing countries wanted to change the currency so as to weaken America). The Americans have helped the Saudis by providing arms and by helping to keep the ruling Saudi royals on the throne (against the wishes of the Saudi people).

In 1953, a CIA coup which put the Shah in power gave Iran to the

United States of America. (The Americans also helped the Shah form his much hated secret police.) And within a couple of years after that Iraq was jointly controlled by America and Britain.

In 1955 America set up the Baghdad Pact, which was designed, at least in part, to oppose the rise of Arab liberation movements in the Middle East. Britain and Iraq were signatories, although Iraq was independent only in name. The British still had military airfields in Iraq, which was ruled by a corrupt monarchy. The people of Iraq, despite having a huge quantity of the world's oil under their feet, were still starving and living in abject poverty.

Things changed in Iraq in 1958. A military rebellion launched a revolution which was to have dramatic consequences for the world. The day after the revolution started the Americans put 20,000 marines into Lebanon and over 6,000 British paratroopers dropped into Jordan. Under Eisenhower's leadership the USA and the UK had made it clear that they would go to war to protect their interests in Lebanon and Jordan.

The British, rather naively, thought that they were simply protecting their interests outside Iraq. The Americans had bigger thoughts. They wanted to go into Iraq, overturn the revolution and put a new puppet government (friendly to the USA, of course) in charge in Baghdad.

But the Americans were stopped. The Iraq revolution was too big. And it had too much support from other Arab countries, from the People's Republic of China and from the USSR. The Americans glumly gave up their imperialist plans.

But they didn't give up permanently.

The Americans added Iraq to their growing list of terrorist nations and gave great support to right wing Kurdish elements who were fighting the Iraqi Government. Then, in the late 1970s the Americans supported the government of Saddam Hussein in its fight against communism. In the

1980s the Americans supported (with money and arms) Saddam Hussein's Iraq in its eight year war against Iran, a country over which America had lost control during Iran's Islamic Revolution of 1979. The Americans openly admitted that they were intervening in order to safeguard their access to the region's oil and they slightly less openly hoped that Iraq and Iran would weaken one another and enable the USA to take over. `I hope they kill each other,' former Secretary of State Henry Kissinger is said to have remarked. The Americans provided Iraq's air force with satellite photographs of Iranian targets and sent anti-aircraft missiles to Iran so that the Iranians could shoot down the aircraft which the Iraqis sent over. America was fighting on both sides in this war and was well aware that Saddam Hussein was using chemical weapons. Over a million people died and both countries were left much weaker. (Bizarrely, and hypocritically, in 2003 George W.Bush, claimed that Saddam Hussein's use of chemical weapons in this war was one of the main reasons for attacking Iraq.) The money America made from selling missiles to Iran was used to finance the Contras who were fighting the socialist government in Nicaragua. Reagan, USA President at the time, disapproved of socialist regimes and wanted to get rid of this one in particular. (It is perhaps unfair to ascribe such depth of feeling to Reagan himself, rather than to his advisors.)

The war between Iraq and Iran didn't finish until 1988, by which time Iraq had become friendly with the USSR.

But then the USSR was taken over by Gorbachev, who wanted an end to the cold war and a permanent detente with America. Gorbachev withdrew Soviet support from Iraq (as he had withdrawn it from countries in Eastern Europe) and the world suddenly changed yet again.

After the war with Iran, Saddam Hussein had accumulated massive debts. The low price of oil meant that his income didn't match his national outgoings. The Iraqi president accused Kuwait of drilling for oil in Iraqi territory and then announced that Kuwait wasn't a separate nation at all but was a province of Iraq. Iraq troops invaded Kuwait in 1990. America (with an international force) attacked, the resultant war was over in weeks and in 1991 the Americans got back into Iraq.

In the decade that followed they used sanctions, bombings and blockades to weaken the Iraqi people and to destroy their spirit. American sanctions against Iraq did not target Saddam Hussein, they targeted the Iraqi people.

When the Americans attacked Iraq in the Gulf War they deliberately bombed the country's water supplies. Then, after the `end' of the war the USA helped ensure that new water purification systems could not be imported into Iraq.

The result was that thousands of innocent Iraqis (including young children) died. The United Nations estimates that more than over a million citizens died as a direct result of the sanctions against Iraq and that unclean water was a major contributor to these deaths. A UNICEF study done in 1999 showed that USA led sanctions on Iraq had resulted in the deaths of 500,000 children under the age of five.

The American Pentagon knew of, and monitored, the destruction of Iraq's water supplies, despite the fact that the destruction of civilian infrastructures which are essential for health and welfare is in direct violation of the Geneva Convention.

The American Government knew that bacteria develop in unpurified water, that epidemics would occur, that the manufacture of safe medicine would be compromised, that food supplies would be affected and that, as a result, there would be thousands of civilian deaths.

When an interviewer questioned the American Secretary of State, Madeleine Albright, about the fact that her Government's sanctions had resulted in the deaths of half a million children, Albright responded: `We think the price is worth it.'

`We have 50% of the world's wealth but only 6.3% of its population,' said George F.Kennan, American Ambassador to Moscow, author of an American State Department Policy Planning Study after World War II. `In this situation, our real job in the coming period is to devise a pattern of relationships which permit us to maintain this disparity. To do so, we have to dispense with all sentimentality...we should cease thinking about human rights, the raising of living standards and democratisation.'

Kennan's paper has been the blueprint for American foreign policy for the last half a century.

Since the disaster of the first invasion of Iraq in 1991 the Americans have been trying to get control of the Iraqi oil. They decided they had to invade when the Chinese and the French did oil deals that would have clicked into place when the sanctions ended.

The Americans knew that Saddam Hussein was no threat to America and that he has no weapons of mass destruction. They also knew that Saddam Hussein has nothing in common with Osama bin Laden.

In 2003 America invaded Iraq for the same, good old reason: oil. By early 2007 the allies had spent half a trillion dollars destroying Iraq's infrastructure and hundreds of thousands had died in the war.

As an aside it is worth noting that the Pentagon is the biggest single user of oil in the world. Tanks, aeroplanes and aircraft carriers are not designed to be fuel efficient and with so many wars going on the American military are burning up oil as if they were trying to get rid of a surplus. As the oil crisis develops (and becomes more obvious) so the military in the USA (and, indeed, everywhere) will stake a very firm claim on what is left. The result must be that the commercial price (the price you and I will have to pay) will rocket ever higher.

Iraq possesses around 11% of the world's oil reserves. I don't think there is anyone left who doesn't now believe that America and the United Kingdom started a war against Iraq in order to snatch control of the oil.

There have, of course, never been any signs that Britain, despite sharing the world's opprobrium for taking part in an entirely unjustified attack on another country, would ever receive any of the oil.

But will America ever actually manage to control the oil it has fought so hard to obtain?

It doesn't look very likely. There have been literally thousands of attacks on pipelines and refineries in Iraq. It seems very likely that the Iraqi resistance fighters will continue to make it difficult for America to steal their country's oil.

(There have, of course, also been many attacks on oil installations in other countries including Nigeria, Iran, Russia, Pakistan, Chechnya and Azerbaijan. These attacks have been designed to disrupt the easy flow of oil to America in particular and the West in general.)

It was always clear (even before the invasion) that America was going to struggle to control Iraq and its oil.

`(USA Policy) is clearly...motivated by George W Bush's desire to please the arms and oil industries,'said Nelson Mandela.

America has successfully demonised any country which has oil and which it does not control. Demonising such countries makes it much easier to invade them without incurring too much displeasure from the American people.

The USA spends vast amounts of money on its army, navy and air force. The American budget gives its greatest priority to the military and under George W.Bush the annual increase in spending on bombs, jets, tanks and guns has been greater than the entire military budget of any other country in the world except Russia. In 2007, America was spending around $1,000 per person on arms. Only Israel spent more.

America claims to be the world's policeman, cracking down on terrorism and totalitarianism, fascism and dictators everywhere. Their aim, say American leaders, is to defend freedom.

This is, of course, a cynical lie. America has shown no interest in countries such as Zimbabwe where millions have died under cruel dictatorships but where there is no oil to be had. America only cares about countries which have oil and its late twentieth and early twenty first century military excursions have been designed with the aim of grabbing whatever resources may be available.

Today, the average American uses five times as much energy as the average citizen elsewhere. Without American greed the fossil fuel crisis would not have hit us for generations to come.

Since the end of the Second World War (which America joined belatedly and only then because it saw huge opportunities for financial and political gain) America has bombed or invaded at least 19 countries and has engaged in direct or indirect military action in many more.

Back in 1980 the Carter Doctrine stated that attempts to disrupt the flow of Persian Gulf oil would be regarded as an `assault on the vital interests of the United States' and would be `repelled by any means necessary, including military force'. Since then America has taken a close interest in Middle Eastern affairs. (What possible other reason could the USA have for taking so much interest in the Arab countries, other than the fact that 60% of proven global oil reserves are there?)

Stealing natural resources in this way may provide America with a fix but it won't change what will happen in the long run. The world is running out of fossil fuels and although stealing what's left from poor countries is clearly wrong and unfair to the citizens of those countries America is merely delaying the inevitable and increasing its dependence on a `drug' that is disappearing.

The danger, of course, is that other countries will follow America's example. (In one way they already have. Countries such as China point to America when they refuse to cut back their consumption of oil.)

America claims to have invaded Iraq in order to impose American democracy on the people there. How curious then that America seems extremely happy with the state of affairs in Saudi Arabia, where a massive 25% of the Saudi GDP goes towards the support of the royal family and where a secret poll showed that half of the population supports Osama Bin Laden.

Saudi Arabia is one of the most repressive states on earth, with no freedom of expression and discrimination against women. And yet America and Britain, who claimed to be horrified by discrimination against women in Afghanistan and Iraq, were perfectly happy to support and defend the despotic rulers in Saudi Arabia.

Justice in Saudi Arabia consists of limbs being amputated and public executions. Defendants have very little right to defend themselves. But the American and British Governments fall over backwards to avoid upsetting the people who are in control because Saudi Arabia is a major source of oil and in the past its rulers have invariably opened up the taps whenever supplies seem to be running a little low.

It's difficult to avoid the conclusion that America imposes its own rather bizarre version of democracy only where it sees that there is a financial or political advantage to be won.

In the 1980s USA President Reagan and UK Prime Minister Margaret Thatcher persuaded the Saudis to increase their oil production in order to bring the oil price down from $30 a barrel to $10 a barrel.

(This was rather stupid of Thatcher and didn't do Britain any favours. As a net exporter of oil it meant that Britain lost huge amounts of money by selling oil at a third of the price.)

The aim was to destroy the Soviet Union, which was dependent on oil exports, and it worked - resulting in the collapse of the Soviet Union in 1991.

The collapse of the Soviet Union was something of a shock for the Americans who then realised that without an obvious enemy they no longer had a bogeyman against whom to protect the American people and the world at large. (And, therefore, didn't have much of an excuse to keep stockpiling weapons and invading smaller countries.)

When Reagan was replaced by George H.W.Bush (Bush the elder) the Americans decided they wanted the price to go up again because American oil companies were suffering. (The Americans never actually think these things through.)

And so the oil price was allowed to rise again.

In the 1990s the Americans eventually realised how vulnerable they were to foreign oil producing countries. The Americans decided not only to increase their presence and influence in the Middle East but also to import oil from as many non Arab states as they could. They used the World Bank, the International Monetary Fund and other organisations to pay for oil explorations and pipelines in Africa, Asia and South America, and to obtain non OPEC oil suppliers.

This complex web of international oil supplies enabled a new company called Enron to thrive. (Enron gave money to politicians inside and outside America in order to seal their supply sources.)

It seems that the formerly massive but now defunct Enron (at one point allegedly the world's largest company - though very few people had ever heard of it until it collapsed and very few people seemed able to describe exactly what it did) gave vast amounts of boodle to 71 out of

America's 100 senators. The company also threw money at George W. Bush during his election campaign.

It has been alleged that Enron's extensive interests in the oil industry meant that the company wasn't terribly keen on America sticking to the Kyoto Treaty. It is common knowledge that one of the first actions of George W Bush, when becoming president of the USA, was to reject the Kyoto Treaty. Could there possibly be any link between these facts?

Enron bought Bush (and America) quite cheaply but the company got the British Government for a much smaller price.

`Oil is too important to be left to the Arabs,' said Henry Kissinger. And you know he meant it.

American military action in the Balkans in the 1990s was undoubtedly motivated not by any desire to liberate the local population but by a search for energy.

The Balkans aren't resource rich but the region is important for moving energy from Central Asia to Europe and thence to America.

The American base in Kosovo, on farmland seized by America, is the largest American military base built since the Vietnam War. Coincidentally, the base is built right next to the Trans-Balkan oil pipeline.

Despite being financially and politically committed to the EU, Britain turned its back on its European allies, severing many of its ties with France, Germany and Italy and allied itself with the USA.

The USA and Britain wanted to ensure the dominance of their defence contractors and oil companies and to establish control over strategic pipelines through and from the Balkans, Eastern Europe and the former USSR.

At one point the American Government is claimed to have deliberately destabilised Macedonia in order to allow easier access for an oil pipeline jointly owned by the USA and the UK.

In Yugoslavia the Americans (with New Labour support from the UK) managed to renew violence between ethnic groups, to provoke a humanitarian catastrophe and to destabilise the Balkans.

The 11/9 attack on America was blamed on a Saudia Arabian called Osama bin Laden. (Osama bin Laden is a member of a family which had long-standing ties to the Bush family - the bin Ladens had indirectly helped George W Bush with his first business venture). Much to the embarrassment of the Americans it turned out that bin Laden had been sponsored by the Americans throughout much of the 1980s. Osama bin Ladens secret bases in Afghanistan were planned or built for him by the Central Intelligence Agency during the 1980s.

It is widely believed that the American and British war on Afghanistan was a result of the 11/9 attack on America. But a French book called `Bin Laden: La Verite Interdite', written by French intelligence analysts Jean-Charles Brisard and Guillaume Dasquie, claims that the Bush Administration in the USA halted investigations into terrorist activities related to the bin Laden family and began planning for a war against Afghanistan before the events of 11.9.2001.

The two authors allege that, under the influence of American oil companies, George W Bush and friends stopped investigations into terrorism while bargaining with the Taliban in Afghanistan to give them Osama bin Laden in return for political recognition and economic aid. It is claimed that the USA Government wanted to deal with the Taliban (rather than overthrow it) so that it could gain access to the oil and gas reserves in Central Asia and build an oil pipeline.

It seems clear that the American inspired attack on Afghanistan was planned for months before the 11/9 attack. Threats of an American military attack were allegedly made to Taliban representatives when the Americans were negotiating the building of a gas pipeline through

Afghanistan to ports in Pakistan. The Taliban Ambassador to Pakistan was allegedly told by an American Government representative that `either you accept our offer of a carpet of gold, or we bury you under a carpet of bombs'. This was in August 2001.

Afghanistan is situated close to significant oil and gas reserves in the Caspian Sea.

Shortly after America started its war against Afghanistan, agreements were signed for the pipeline through that country.

It has also been alleged that the USA had been planning to invade Afghanistan for as long as three years before the 11/9 attack. It has been reported that the USA Government told the Indian Government, in June 2001, that there would be an invasion of Afghanistan in October 2001. Defence analysts had reported the planned invasion as early as March 2001.

After the infamous 11/9 attack on America George W Bush announced that its war on Afghanistan was just the beginning of its `war on terrorism'. Bush made his infamous `you are either with us or against us' speech and a list of nearly 50 target nations was published. Most of the nations on the list had important oil resources but had no links with bin Laden or Al Qaeda.

After studying details of the 11/9 attack on America many independent observers believed that the attack was inspired, orchestrated and possibly even carried out by the American government itself as an excuse for grabbing control of the world's oil reserves. There is no doubt that the so called `war on terror' could be more accurately entitled the `war for oil'.

George W.Bush's backers, the American neoconservative zionists, saw what was happening some years ago. They have, therefore, tried to create a world in which they will control what oil exists, benefit from the shortage of oil and be free to introduce an endless variety of legislation designed to limit our freedom and expand their power.

The legislation which has changed the world since 11th September 2001 was clearly brought in to enable a relatively small number of money and power hungry men (and women) to control the world and to control potential rioters.

Today, wherever there are significant oil or gas pipelines or fields there will be an American base nearby. The only two significant exceptions are Russia and Iran. American troops are stationed in 120 countries.

American oil companies pay the Islamic Government in Northern Sudan so that they can gain access to untapped oilfields there. And American Christian groups finance the non-Islamic Southerners because they believe that in doing this they are helping to fight the war against Islam. The result: civil war, paid for pretty much entirely by Americans.

The Americans would have loved to invade Iran (and it was widely rumoured that they were planning to do so in Spring of 2007). They had certainly been looking for excuses for an invasion.

In the end they didn't invade for purely practical reasons: they didn't have enough men left (the wars against Iraq and Afghanistan had both proved more troublesome than expected), they didn't have enough money left (America is effectively bankrupt and wars are very expensive) and they were frightened of China (which had formed a close alliance with Iran).

It is important to understand that the modern American version of Christianity seems to allow politicians to pick and choose the dictators they attack. They go for the ones who have oil or who won't do business with us but remain on good terms with the ones (in countries such as China, Zimbabwe) with whom profitable relationships have been established. The Chinese Government is no better than Saddam Hussein's Government but the Americans would never dream of invading China. For one thing their currency depends on Chinese support. And for another they know that they would lose a war with China. America, like all bullies, only tackles weaker targets.

The war on Iraq has been an unmitigated disaster. Thousands of American and British servicemen and women have been killed. It is difficult to know how many Iraqi civilians have been killed (neither the Americans or the British bother to keep count of Iraqis who are killed) but independent observers put the figure at around a million. After three years of the war The Lancet has reported that the death toll in Iraq exceeded 650,000. This puts George W.Bush and Tony Blair high up on any list of the worst war criminals of all time.

During the run up to the 2003 Iraq War the Americans, desperate to get the Russian vote for the United Nations security council resolution that would give them the green light to bomb Iraq and grab its oil, promised the Russians that Iraq's outstanding $8 billion debts to Moscow and the Russian oil industry would be honoured in a post-Saddam Hussein Iraq.

In principle the Americans did not, of course, have any right to make decisions for a post Saddam Hussein regime in Iraq. In practice the Russians presumably knew that America would, as conquerors, have control of Iraq's oil and its money. It was, perhaps, a sign of the Americans' desperation that they were prepared to make this deal and, thereby, be rather more obvious about their intentions than they had previously been.

China and the USA are at a standoff over Iranian oil. China, which now has most of the money in the world, has for years been wooing the Arabs. They have offered to support Iran if the Americans should invade.

Iran has the second biggest oil reserves in the world and recently agreed a $70 billion 25 year deal to supply oil to China.

As the oil runs out there are bound to be more wars over the diminishing quantities of fossil fuels left on the planet

There have always been wars over resources.

Men have fought over everything of value, but resources such as land, horses, cattle, ports and waterways have always been top of the list. As the oil runs out so the wars are likely to be more violent, more common and more desperate.

America is in decline. Its place as the world's controlling nation has been a short and violent one.

Since the Second World War, American foreign policy has been dictated by its yearning for oil. America's best move was persuading the Arabs to sell oil in dollars. This has meant that every oil importing country in the world has had to pay for oil in America's currency. It is to a large extent through this piece of financial trickery that America has built up huge debts and has nevertheless apparently remained rich.

When will the remaining oil producing nations insist on selling oil in euros instead of dollars?

Despite its yearning for Arab oil America has continued to defend Israel unquestioningly. Powerful Zionists in American politics are doubtless partly responsible for this. But America has also used Israel as a local staging post; enabling them to keep an eye on what is happening in the rest of the Middle East.

America now regards anything the Palestinians do as terrorism. In contrast, anything Israel does is regarded as self defence. The media has helped create and defend this myth.

Israel also helps to act as a focal point for Arab resentment - taking a little of the heat off America.

These policies are now falling apart, of course.

America's violence against Arab countries has resulted in so much hatred against the USA that it doubtful if the rulers in Saudi Arabia can remain in power much longer. There are some who feel that America invaded Iraq so as to be close to Saudi Arabia when the bin Laden supporters there eventually overthrow the Saudi royal family. Losing access to the Saudi oil would damage the USA enormously.

Around the rest of the world America has made enemies almost everywhere. Although there may be fragile economic links between America and China the reality is that there are huge divisions between the two and neither country trusts the other. The same goes for Russia. Much to America's horror both China and Russia have both developed close links to Iran.

`Twenty years on from the oil shock of the 1970s, most economists would agree that oil is no longer the most important commodity in the world economy,'said Tony Blair, the British Prime Minister, in January 2000. I have found no evidence that Mr Blair ever explained what he thought the most important commodity might be.

It is difficult to know just why the British Government allied Britain so closely with America. Tony Blair, the Prime Minister at the time, put forward several reasons for taking Britain to war against Afghanistan and Iraq but none of them have much of a ring of truth. And Blair's credibility was of course so damaged that for almost the whole of his premiership it was difficult to believe anything he said.

The most generous thought is that Blair realised that with the oil and coal running out Britain would have to find new sources of energy. (Though this thought seems unlikely since Blair famously - and rather stupidly - claimed that the new information economy had replaced the oil economy.) If he did think this then his policy was a total failure since there are absolutely no signs that Britain will receive any of the oil that the Americans have now managed to steal.

My own suspicion, I fear, is that Blair was simply behaving as Bush's poodle so that he could rely on Bush and America for lucrative employment once his stint as Prime Minister was over.

Is there any doubt that America will eventually turn against Europe and use whatever military might it has left to grab whatever resources might be available?

Of course not.

America, a nation founded on slavery and genocide, has always looked after America first and recent administrations have proved themselves to be corrupt and untrustworthy in the extreme.

Chapter Four: What will happen when the oil runs out

The oil won't run out overnight.

The oil isn't going to be there on Tuesday and gone on Wednesday.

But it will become scarcer.

And every time there is a calamity of some kind the price will go up significantly.

When a hurricane destroys an oil rig off the American coast the global price will go up. Noticeably. When Hurricanes Rita and Katrina hit in 2005 oil production was reduced by a relatively small amount - just 1.5 million barrels per day (bpd). But that was enough to push oil prices up. Afterwards, oil prices went back down again. In the future oil prices will hardly ever go back down again.

When there is a cold spell in the USA, and Americans turn up their heating for a week the price of oil will go up. Noticeably.

When there is a problem affecting oil exports from Iran or Venezuela the price will go up. When there is a problem with a production in Saudi Arabia or Kuwait the price will go up. If terrorists manage to blow up a pipeline then prices will go up.

And the price will stay up.

It will slowly become clear that the world has no spare oil capacity. Every disruption will emphasise the truth.

`Sooner or later, we sit down to a banquet of consequences,' wrote Robert Louis Stevenson in 1885.

Remember, while the oil runs out, America, the UK and other oil importing nations will still be vulnerable to terrorism, an Islamist revolution in large oil producing nations, or the whims of Russia.

And it isn't just oil that is at risk. At least 60% of the world's gas is concentrated in Russia, Iran and Qatar. These are hardly the most stable nations on the planet.

And remember, too, that while our relations with the Arab world can hardly be said to be `friendly' the Russian energy business is under the ruthless control of crooks and former KGB men.

The oil importing nations depend for their future supplies entirely on continuing to have good relations with (and there being stability in) Nigeria, Iran, Russia and Venezuela.

Nations which have oil will gradually become increasingly aware of their power. And increasing possessive about their supplies.

When oil prices rocket (and they will rocket because as soon as it becomes clear that the oil supply is threatened - and long before it runs out - the world's military will grab what they can and politicians will start stockpiling what they can store) there will be chaos.

Unlike America (which has several years supply of oil in huge caverns) Britain has no more than a few days supply of oil stored. And without oil everything will be in short supply.

What will make the oil crash more damaging and more destructive is the fact that the global economy is already overheated, overstretched and therefore weak and unhealthy.

Thanks to the European Union and the compliant feebleness of successive governments, many countries (particularly Britain) have farming industries which are in tatters.

Governments have allowed their farming industries to be destroyed by the import of cheap crops from abroad. There will, therefore, be a shortage of home grown food.

Add to this the fact that for every calorie of food we eat about ten calories of fossil energy are used to bring it to our plates.

So when the cost of petrol rockets so will the price of food.

The rising price, and increasing shortage, of fuel means that our whole future will have to change.

Bicycles and, for the rich, horse drawn traps, are the future. City taxis will be replaced by rickshaws. Towns with a good connection to the rail network will survive. Others will die.

A post oil world will be one of rampant inflation, recession and wars wherever there is oil.

Many Western countries already have fascist governments. Governments will respond to these new dangers and threats by announcing that they need to take even more power.

For the protection of the public, of course.

`It has been a fabulous party,' wrote Richard Heinberg, `The Party's Over' in 2005. `But from those to whom much has been given, much should be expected. Once we are aware of the choice, it is up to us to decide: shall we vainly continue revelling until the bitter end, and take most of the rest of the world down with us? Or shall we acknowledge that the party is over, clean up after ourselves, and make way for those who will come after us?'

High inflation rates mean that savings are eroded.

If inflation rates exceed interest rates (as can, and will happen) cash in the bank or building society loses its purchasing power. Investments in shares will be a disaster too. As companies struggle to survive so dividends will dry up and share prices will collapse. Anyone who tries to survive by buying an index fund, and trusting that the market as a whole will survive, will lose money. Investors who have grown accustomed to the idea of buying a diversified portfolio and sitting back to wait for profits to come will suffer great pain as they see their savings melt away.

As the oil price goes up the problems of our world will, for a while, resemble the temporary problems of the 1970s. High inflation, high interest rates, high unemployment and falling house prices will, I believe, combine to create great misery among those who are not properly prepared.

Investment policies which worked well in the 1990s or the early part of the twenty first century will fail miserably. Savings, nest eggs and pension plans will evaporate.

In the past government employees have always been pretty well immune to the horrors of a collapsing economy. They have considered that their jobs and salaries are safe, as are their pensions.

But this will change.

As the oil price rises, rises and rises again and as the economy falters, falters and starts to collapse the Government's income will no longer be sufficient to pay its bills.

Governments will, of course, raise taxes. Even though services will continue to decline income taxes, corporation taxes, capital gains taxes, national insurance, VAT and stealth taxes will all go up.

But there will come a point when taxes cannot rise any more.

And then the unthinkable will happen.

The Government will have to start laying off vast numbers of civil servants. And then it will find itself unable to continue to pay absurdly over-generous pensions to the civil servants who are already retired.

For a while the Government will try to borrow enough money to fulfil its obligations. But this won't work. Where can a bankrupt Government borrow money? If a government tries to print its way out of trouble the currency will collapse. Countries everywhere will resemble Germany in the 1920s.

Russia has, for many years, been one of the world's most significant sources of oil. At the start of the 20th century Russia was for a while the world's largest oil producer. Only the discovery of huge oil fields in the USA pushed Russia into second place. But will Russia continue to supply the west with its oil?

Possibly not.

For one thing Russia needs its oil supplies for itself.

And for another, the Russians know that during the 1980s the Reagan Administration in the USA persuaded the Saudis to flood the world with cheap oil in order to undercut Russian oil and to bankrupt the former Soviet Union.

Does anyone in America really believe that when the tables are turned (as they will be soon) the Russian Government will fall over backwards to rescue America or its allies?

Nations which don't have enough oil will fight to get what they want. At a time when nations should be running down their military machines (all those tanks and aeroplanes use vast amounts of fuel) the very opposite will happen. Valuable oil supplies will be grabbed by the military.

Will China and America fight for the planet's remaining oil? If it does then I wouldn't put your money on America. The Chinese have the manpower and the will to win.

Will there be a nuclear war?

There are those who believe that the chaos which followed the American invasion of Iraq was deliberate American policy. The chaos and lawlessness enabled America to create a government and gave Bush an excuse to keep his troops in the country. The rioting and civil disobedience also enabled the USA to draw Iran and Syria into the conflict.

Was this deliberate regional destabilisation?

`What keeps us awake at night is the threat of the growing energy crisis that, if our civilisation does not deal with it in time, will affect every community on earth and last forever,' wrote Stephen Leeb in `The Coming Economic Collapse'.

In his book `The Long Emergency', James Howard Kunstler explains what will happen when oil supplies decline. `We will be compelled by the circumstances of the Long Emergency,' he writes, `to conduct the activities of daily life on a smaller scale, whether we like it or not...'

The title of Kunstler's book says it all.

`The Long Emergency'.

This is one emergency that isn't going to go away.

And slowly we will have to become accustomed to the idea that our world is changing.

Rising petrol prices will mean that using motor cars will become a luxury. Private vehicles are an absurd self indulgence.   Just think of it: all those vehicles on the motorway. Most of them containing one or two passengers. That will have to stop. Bicycles and the horse and cart will be widely used.

On the railways, diesel engines will be replaced by old fashioned steam trains. The railway museums will be emptied of legendary steam locomotives.

Horse drawn carriages will make a come back and people will start experimenting with land yachts.

Goods will be transported around the world on old fashioned clippers.

At the moment 90% of all goods are moved around the world by 50,000 diesel powered container ships.

How else are we going to move things around the world without oil?

In fact, it's already happening. In the summer of 2007 the Germany shipping firm, Beluga, announced that it was launching a container ship with a giant kite flying 1,000 feet above the bow. The 1,100 square foot kite was connected to an automated telescoping mast. The wind power produced was estimated to cut fuel consumption by as much as 30%.

(This isn't the first time a modern shipping company has considered wind power. In the 1980s, after a spike in oil prices, Japanese shippers experimented with wind assisted vessels.)

Container ships, passenger cruise ships, fishing trawlers, oil tankers, big yachts - are all likely to be wind powered. Ships and boats which rely exclusively on oil to move around will be scrapped.

'We are all addicts of fossil fuels in a state of denial, about to face cold turkey,' wrote Kurt Vonnegut.'And like so many addicts about to face cold turkey, our leaders are now committing violent crimes to get what little is left of what we're hooked on.'

No one will want big houses because the heating costs will be too high.

The increasing cost of using tractors and other farm machinery will add to the cost of food prices.

The shortage of oil will mean that fertilisers and pesticides become more expensive too.

Transporting food around the world will become impossible. Local markets, selling produce grown locally, will boom but food will become increasingly scarce and, therefore, increasingly expensive.

(If governments around the world continue with the madness of encouraging the use of ethanol then food will become prohibitively expensive for millions.)

Supermarket shelves will not be restocked quite so quickly as they used to be. Within a year or two of the oil crisis hitting supermarkets will close.

All shops which rely on selling products imported from other countries will struggle. Thieves will be stealing tins of beans in preference to television sets.

The cost of raw materials other than oil will also soar. Digging metals out of the ground uses up a great deal of energy. And so zinc, copper and other metals will rise in price. The cost of steel will go up. Wood will continue to be scarce and expensive. Wood from old buildings will be salvaged not burnt. Building a house will cost more but as unemployment rises so house prices will fall. The result will be that builders will go out of business and there will be a shortage of homes.

Car manufacturing will become prohibitively expensive (particularly since very few people will be able to afford petrol). People who continue to drive will have to mend their old cars instead of buying new ones. Car show rooms will, like other retailers, close down.

Airlines will no longer be able to offer cheap flights. Holidays abroad will be a thing of the past for most people. Travel agents will close down. Whole industries have no long term future at all. The advertising industry, in all its forms, will die and few will mourn it.

Globalisation, in all its forms and all its meanings, will become a footnote in history books. Those who hoped for a world government will have to abandon their dreams. It won't happen.

Some countries will survive better than others. France, which obtains three quarters of its electricity from nuclear power and which has (thanks to the European Union's Common Agricultural Policy) a huge, relatively old fashioned but still successful farming industry, will survive quite well - though it will have to get rid of most of its civil servants. France will be one of the favourite targets for emigrants leaving Britain and America.

Britain will be among the countries most likely to flounder. The Labour Government has created an untrusting, unworkable form of socialism which will not survive. The debt ridden nation they have built will suffer more than any other developed nation on the planet - except America.

For financial and political reasons, America has been doomed for a decade or more, but the peak oil crisis will cause turmoil and will leave America impoverished and in chaos.

The American way of life depends on a constant supply of cheap oil.

France, Germany, Spain and most countries in Europe have extensive railway services. And, despite the egregiously stupid and short-sighted activities of the ludicrous Beeching Plan, most decent sized towns in Britain still have a railway station of their own. It is possible to travel around Europe by train without too much inconvenience. (Homes within walking or cycling distance of railway stations will be the few to go up in price.)

America doesn't have a nationwide railway network. The success of road transport did enormous damage to rail transport in many countries but particularly in America. As a result, America is in a terrible position as far as public transport is concerned. The International Railway Journal puts America between Bolivia and Turkey for the amount spent per head of population on mainline railway expenditure. And yet good railways are relatively cheap. Switzerland, which probably has the best railway system in the world, spends a little over £100 per person per year on maintaining its railway.

Roads everywhere are built by governments. Rail tracks have to be built and maintained by the railway companies so, unless governments helped the rail companies, the trains never stood a chance.

In America today people either fly or drive.

And flying and driving are going to be increasingly expensive and eventually impossible for ordinary people. The number of private jet owners in the USA has doubled in the last ten years to more than 50,000. Celebrities now use their private planes to fly around the world in oil-guzzling luxury. (Many, with no sense of irony, and unconscious of their hypocrisy, preach about the dangers of global warming and exhort their fans to eschew their annual holiday in Spain in order to reduce their carbon footprints.)

America is also vulnerable because it has relatively little energy from nuclear power. It relies far too heavily on oil and coal for its power supplies.

America faces another problem which will make things worse. The country is deeply in debt. Even without the oil running out America would be heading for extremely serious economic problems - an economic depression that would make the 1930s look like a long holiday.

There will be a meltdown of the American dollar that will be reminiscent of the collapse of the German mark during the Weimar Republic in the 1920s.

Things are not helped by the fact that other countries (particularly China) are now huge holders of American debt.   Foreign holdings of American securities have doubled since 2002, mainly because China (which now owns trillions of dollars) wants to keep the dollar high so as to support its own exports. Around 80% of USA Government bonds due to be redeemed in next 3 to 10 years are held by foreigners.

This can, and will, change in days.

The Coming Great Depression will lead to the end of Medicare and Social Security in the USA. America simply will not be able to afford these programmes.

The disintegration of American society will lead to civil unrest, to rioting and to enormous amounts of violence. Thanks to the powerful pro-gun lobby in the USA a lot of the unhappy citizens will be armed. (There are enough guns in America for every man, woman and child to have one each.)

Everyone will be looking for someone to blame. The liberals will want to blame the Bush family and the zionist neocons. The republicans will look abroad and will want to blame bin Laden. Extreme left wing groups and environmentalists will blame big business. Everyone will be too busy blaming someone else to do anything. The rioting will lead to lynching. America will break up into separate states.

The end of cheap oil will lead directly to the end of America as it exists today.

It is worth repeating (and remembering) that the collapse of the German economy in the 1920s and 1930s led to the rise of Hitler and the Third Reich. It is impossible to predict just what will arise out of the ashes of America. But it probably won't be very pleasant.

The USA was successful and strong throughout the early and middle decades of the twentieth century because of its huge natural resources - especially oil. (It was successful throughout the later decades of that century through greed and aggression.)

Access to its own supplies of natural resources allowed America to take advantage of European inventions such as the train, the aeroplane and the motor car.

Now those valuable home grown resources are running out. The oil has pretty much gone. And America is on the slide.

The American transport system is 97% dependent on oil and more than 90% of the oil supply is controlled by foreign governments.

Things will get so bad in America that it won't be long before the Americans are smuggling themselves across the border into Mexico in search of jobs. They certainly won't be able to buy any oil unless the Iranians suddenly start accepting buttons as currency.

`Most likely the end of the Petroleum Interval will be gradual wherein no crisis point is reached, just slow change. But, especially with continually rising populations, and no sufficient substitutes for oil at hand, there is the possibility of a chaotic breakdown of society,' wrote Walter Youngquist in `Geodestinies: The Inevitable Control of Earth Resources over Nations and Individuals'.

So, since we are now almost certainly on the downward slope why haven't currencies started to unwind already?

Why isn't the American dollar in an even steeper decline than it is?

There are several reasons.

First, not many people realise the truth that you now know. The truth about peak oil is something that governments have done their best to hide. And the significance of peak oil is something that they have hidden very well. We have been persuaded to worry about global warming rather than peak oil. We have been told that we must cut our use of oil because of global warming - not because the stuff is running out.

Most people still hope and believe that the oil will continue to flow. Even sophisticated investors believe that the oil companies will simply find more oil fields to replace the ones which are dying. Many investors who are aware of the reality of peak oil simply find the prospect too frightening to accept. They prefer to stick their heads in the empty oil wells and ignore what is happening.

And since people believe that there is no real oil problem they also believe that there is no threat to growth.

And because they believe that there is no threat to growth they also believe that the bits of paper they are holding (share certificates and bank deposits) will continue to hold their value.

The second crucial reason why the American dollar has not yet collapsed is that many of the countries to which America owes money are reluctant to allow the dollar to fall. They need to keep the dollar as high as they can because their own wealth is now dependent on America's wealth and financial standing. If America collapses then they too will collapse.

At best, our society will go back to the way it was over a hundred years ago. At worse we will go back to an existence much closer to the Stone Age. Whatever happens, our lives will have to become simpler. Things will never again be as good as they were in the 1960s.

Because it will be more difficult to move around, communities will become more important. Local farms will have to feed local people and will have to employ more farm labourers to replace the equipment they can no longer afford to use. Suburbs will have to become self-reliant. Those which cannot will not survive. The sort of suburban lifestyle many people live today will become impossible. (Kunstler describes the suburban lifestyle as `the great misallocation of resources in the history of the world').

Only the very rich will be able to afford to heat large homes or to own and run a car. Most people who work for a living will have to live within walking or cycling distance of their place of employment. There will be no more imported television sets, Play Station games or out of season foods from the other side of the world. Small, local market gardens will do well.

The collapse of the electricity grids will be the final trigger for our civilisation's decline.

Our lives will, in many ways, become simpler and, perhaps even less stressful. Our needs will be quite obvious, practical and basic.

People will stop eating meat. Farms will start growing crops instead of breeding livestock. The world will turn vegetarian. Only the very rich will still eat meat of any kind.

It will be increasingly difficult to obtain clean water supplies. The collapse of local services will result in sewage disposal problems. Infections will become commoner. Global warming will affect the type of infections to which we are susceptible. Hospitals use an enormous amount of energy and only the wealthy can expect to have access to the most sophisticated hospitals. The development and distribution of drugs will become increasingly difficult. Public health problems will abound unless governments can be persuaded to move money from the military to health care.

The information age will come to an end almost as soon as it has started. The internet will not last long once the oil starts to run out; it will be a hardly mourned memory.

The shortage of oil won't have an immediate effect on information processing but the manufacture of computers, servers and so on will be affected. Once sophisticated systems break down it will not be possible to replace them.

Electricity grids will be subject to frequent interruptions as the supply of energy sources becomes increasingly difficult to sustain. Demand will go up as populations continue to grow, but supply will deteriorate. There will inevitably be shortages. There will be periodic blackouts. Electricity prices will inevitably rise and non-essential usage will halt. The storage and exchange of information electronically will end. The internet will be no more.

As energy supplies become increasingly scarce and expensive so we will become more and more dependent on our neighbours and less and less interested in what is being decided by national or international politicians. National parliaments will lose much of its relevance. The European Parliament will become totally irrelevant and the EU will collapse completely. Regional parliaments, those enormously expensive and secret layers of administration, will wither and die. Vastly reduced incomes will force local councils to slim down.

Schools will close. Gardens and parks will look unkempt unless the areas of grass are small enough to be cut with a hand mower, or large enough to be controlled by grazing sheep. There will be no more ornamental flower beds.

Every home with a garden will plant vegetables. The streets will be full of the homeless, the disabled, the mentally ill and the deranged.

Trees everywhere will be chopped down for fuel. Those cutting them down will start with trees in the countryside. Then they'll cut down trees in parks and gardens. And eventually trees lining suburban roads will disappear too. Soon a tree will be a rarity.

Office blocks will fall into disuse. No one will want to buy them - or have any use for them. Roads will go unrepaired. Not that it will matter much. There won't be many cars around. Just the bullet proofed limousines of the rich and a few beaten up old jalopies held together with string and sealing wax. Nothing will be as good as it was.

Most universities will disappear. Many jobs, for which people have been extensively trained, simply won't exist. Former computer technicians will find themselves working on the land as farm labourers. Fertilisers, pesticides, herbicides and machinery will be replaced by men, horses and manure.

No one will throw things away when they stop working. Washing machines and cookers will have to be repaired until they fail completely.

Older devices, which contain less complications, will be popular because they will be easier to mend. When the washing machines and cookers go the washing will be done by hand and the cooking done on a stove powered by coal or wood. Clothes will be mended when they wear out. Very few shops will be selling new clothes and very few people will be buying them. The fashion industry will be a footnote in social history books.

There will be a small boom industry in making and selling hand tools. Private colleges will offer courses in practical woodwork, plumbing and household repairs. No one will bother with certificates. Skills not qualifications will be what count. Television sets will not be replaced when they break down. Programmes will mostly be repeats. Computer games will have long since disappeared. (The kids who used to play `end of the world' games will now find themselves living their game lives for real.) And the cult of the celebrity will end. Most people will be too concerned with staying alive to worry about the activities of celebrities - whether minor or major. For most people life will be a question of finding adequate food, clothing and shelter. The sick and the elderly will suffer most.

The Government will no longer be able to afford the vast array of benefits introduced during the last few decades. Taxes will rise higher and higher and services will deteriorate. The absence of any sort of oil (let alone cheap oil) will led to the collapse of what little industry is left in the West.

The British economy, weakened by a decade of fiscal incompetence will collapse.

There will be rioting and revolution as millions of hungry, unemployed people move about the country looking for money and food. For a while, armed police will control the rioting. And then, when the police are no longer paid, there will be no controls and no law.

No government will be able to find enough money to pay all its obligations when things get really tough. Benefits and pensions paid out by the government will either be suspended or slashed dramatically. (Though, I suspect that the architects of our coming disaster, politicians and the more senior civil servants, will find a way to ensure that their pensions are secure.)

Politicians will be too busy looking for someone to blame to take useful action. And, judging by past experience, they will in any case be too aware of the wishes of lobbyists and corporate donors to take much interest in the genuine needs of those in their care. Central government will completely lose touch with the voters. Democracy, already a long distant memory in most countries, will be forgotten. Politicians will attempt to deceive the electorate and will lie about the causes of the problems people are experiencing. Those who do not know the truth will probably be convinced by the lies they are told.

The two countries which will suffer most will be Britain and America.

In Britain, everything that will happen will be made infinitely more difficult by the actions of the Government in the last few years. The closing down of rural post offices, the destroying of small farms, the closing down of rail links to villages and small towns, will all make life infinitely more difficult for everyone.

While all this is happening we will be struggling to cope with the changes produced by global warming. Freak storms and long droughts will destroy crops and make farming much more difficult and much more unreliable.

And that takes us to the biggest problem of all: food.

The cost of food is already going up. Food inflation is now at its highest for years. Wheat, corn and soya prices are all soaring. The price of rice went up 50% between 2005 and 2007. The price of wheat went up 33% in the same period. And the price of cattle went up by more than 40%. The price of vegetables went up 10.2% between 2006 and 2007. The price of fish went up 12.6% in the same period. The price of onions has rocketed. The price of bananas went up 500%.

And, because food is now a global commodity, these food price changes are a global problem. Early in 2007 a 60% rise in tortilla prices provoked rioting in Mexico where thousands were starving because they could no longer afford to eat their normal diet.

There are several explanations for what has already happened.

First, there has been recent massive rising demand from emerging economies. The people of China and India can now afford to buy food and so they that's exactly what they are doing. Twenty years ago most people in the world subsisted on 1,600 calories a day. Now they want to get fat like the Americans and the British. India produces 70 million tons of wheat a year and is the second largest wheat producer on the planet. But India is now a net importer. Local prices have rose by 12% between 2006 and 2007.

Second, the people of China and India want to eat western foods. They want to eat meat. No longer satisfied with a bowl of rice they want to dine on burgers. The consumption of meat in China will probably double in the next few years. The most protein efficient meat is poultry. A bushel of corn produces 19.6 pounds of retail chicken, 13 pounds of pork and 5 pounds of beef. Sadly, the demand for chicken has been hit hard by the bird flu scares. And so people in China and India want red meat and the consumption of beef in China is growing by 20% a year.

But there are problems. One is that with so much land being used to grow biofuels there is very little land led for growing food for animals. In June 2007 farmers in California were complaining that the cost of a truckload of hay had gone up from $2,000 to $5,000. And turning vegetation into meat is grossly inefficient and costly; it takes 8.3 grams of grain to produce 1 gram of beef.

Third, the world's population is exploding. The world's population is currently around 6 billion. At the rate things are going it will be 9 billion by the year 2020. Many of the extra 3 billion will be in China and India. They won't want to starve. Just to cope with the population growth the world's food production will need to increase by 50%.

Fourth, as populations grow and people want to live in nice suburban houses with neat little lawns. As this happens, so the amount of land available for arable use falls. Every year for the past decade China has lost fertile land equivalent in size to the area of Scotland. To feed its growing population it needs to be increasing its land area by the equivalent of Scotland. Whoops. Things are the same in India.

Fifth, encouraged by politicians, vast quantities of the world's crops of corn, soy bean and so on, are being used to make biofuels so that American motorists can continue to buy cheap petrol for their huge motor cars. Time magazine recently published a list of 51 things you and I can do to prevent global warming. Number 1 on their list was headed `Turn food into fuel'. This, they claimed, would have a `high impact' on the global warming problem. Time magazine claims that ethanol is the alternative fuel that `could finally wean the US from its expensive oil habit and in turn prevent the millions of tons of carbon emissions that go with it.' As I have shown elsewhere in this book this is dangerous nonsense. The International Energy Agency estimates that demands for crops for biofuels will soar from 41.5 million tons of oil equivalent in 2010 to 92.4 million in 2030. (These relatively modest predictions seem to assume that oil will continue to flow and that massive new oil fields will continue to be discovered.) The increased use of biofuel is a major force behind the rise of food prices. Last year 5 billion gallons of ethanol were produced in the USA. Encouraged by George W. Bush this is set to increase to 45 billion gallons by 2009. That, just for the record, is 136% of American corn production. So, either the Americans are going to have stop growing other foods or else they are going to have to import a great deal of corn from around the world. Either way, there is going to be a global shortage of food and millions are going to die. Just how many people will die as a result of the biofuel policy isn't clear and certainly isn't likely to affect American policies. (Iowa, the place where most American corn is grown also happens to be a state which has a big role to play in choosing the next American President. Farmers in Iowa are gloriously happy about the biofuel phenomenon.)

Sixth, there is global warming. Whether you agree with the widely accepted theory that global warming is caused by carbon emissions or not there is little doubt that there have recently been some pretty freakish conditions around the world. The governor of the Bank of England in the UK recently blamed the rise in inflation in Britain on: `a rise in food prices caused by weather-induced global reduction in supply'. The Prime Minister of Australia recently described the drought in his country as `an unprecedentedly dangerous situation'. If things don't change rapidly in Australia then water supplies will dry up and food prices will rocket skywards. The production of rice has already fallen from 1.6 million tons to 106,000 tons. The reality we have to face is that at the same time as our world runs out of oil (making farming an infinitely more difficult business) we will also have to deal with weather conditions - droughts and storms - which also make farming even more difficult.

Seventh, big American seed companies have been busy patenting the rights to many individual seeds. They have done this so that they can force farmers around the world to buy their products. One result has been that small farmers in India are no longer allowed to grow seeds from crops that their families have been planting for generations. (If they do, then lawyers for American multinationals will smother them with writs, injunctions and a typhoon of typical American legobabble.)

Eighth, large modern farms are remarkably (and surprisingly) inefficient. When the fuel used to build tractors, make fertilisers and pesticides and so on is taken into account it turns out that the energy cost of a kilogram of corn has actually risen in the last few decades. Soil erosion, the loss of pollinators (such as bees) who have been killed by chemicals, evolving chemical resistance by pests and numerous other environmental problems have also reduced farm crops.

The result of all this is that food is becoming scarce and prices are rising. This is not a cyclical change (with prices falling next year due to better weather and better crops). It is a structural change. It is, in other words, permanent.

As far as food prices are concerned the conditions are optimum for a `perfect storm'. Things really couldn't get much worse.

(Actually, they could. There is a joker. American genetic engineers have been `modifying' food for years to make it more profitable. No one knows what effect their modifications will have on the safety of food for human consumption. No one knows what other horrendous side effects there might be. The risks are unbelievably dangerous.)

For those in Europe and America all this is no more than a nuisance at the moment.

But for those in many parts of the world this is already an outright disaster. In Guatemala, for example, nearly half of all children are malnourished. And things are getting worse and will continue to get worse. Rising prices and falling quantities of food available for eating (as opposed to filling petrol tanks) will result in massive starvation around the world.

Increasing agricultural production enabled the world to grow from 1.7 billion people to over 6 billion people in just a century. But when the oil runs out the world will not be able to feed that many people.

How many people will the planet feed?

Well, it's a safe guess that it will support around as many people as there were before oil changed farming. So we've got four billion people too many.

Except that during the last century we've done a lot of damage to our soil and our environment.

So the planet probably won't support even one billion people.

If we don't voluntarily reduce the size of the planet (and there are no signs that any nation will choose this route) the answer will be famine, plagues and war.

Five billion people have to die very quickly.

Welcome to your future.

And yet the media continues to ignore this threat.

In the spring of 2007 Fortune magazine carried a feature listing the world's seven biggest problems (and how companies can make money out of them). Their seven problems were (presumably in order of importance):

1. Global warming
2. Hunger and malnutrition
3. Waste disposal
4. Dirty water
5. Dirty air
6. Epidemics
7. Overfishing

The seven problems are all valid.

But Fortune magazine didn't even include Peak Oil on its list.

Maybe because there is no way to deal profitably with the problem.

When the oil is running out and there is no more to find then there cannot be any more to sell.

How will people behave when the oil runs out and the world changes? Will people help one another? Or will there be disorganised chaos, with rioting in the streets and talk of revolution?

It isn't hard to guess.

There has already been a breakdown in law and order all over the Western world. Our security is threatened by Islamist radicals. Our streets are so dangerous than in many cities law-abiding citizens no longer go out after dark. The streets belong to drug gangs and roaming teenagers, drunk and bored. Looting and rioting are the usual response to any sort of short term crisis. Even though many people no longer bother to report criminal attacks (either on their person or their property) crime rates are rocketing. There is a widespread belief that the police are too busy chasing motorists (an easy source of extra finance) to pay much attention to criminals.

When the oil crisis hits our society will fragment. Small but hostile groups will develop, each looking after themselves with no concern for others. The weak will be left to look after themselves.

In a traditional society the family unit would have been the saviour of civilisation. But in many countries the family unit no longer exists.

The family unit is and always has been the basic building block out of which a society is built. For its own political reasons Britain's Labour Government has done its best to destroy the family unit. And it has been enormously successful. Britain's tax and welfare policies provide financial incentives for lone parenthood. Marriage is penalised. Under the Labour party's influence there has been a dramatic growth in the incidence of teenage pregnancy and a rise in the number of single parent families.

Young men father children they have no intention of looking after. Young women deliberately try to get pregnant so that they can be given a home of their own and a weekly fistful of cash to spend. Half of all children are now born outside marriage. The Government has even introduced and promoted homosexual marriages. A government funded group produces a leaflet aimed at 13 year olds which includes advice on `how to be good at sex'. Sex lessons in schools include advice on anal, oral and digital sex for twelve year olds. It is hardly surprising that one in ten women between 16 and 25 is now affected with chlamydia - a sexually transmitted disease which can cause infertility. The deliberate destruction of the family will mean that the elderly, the disabled and the weak will be left to die.

Extended drinking hours (brought in courtesy of the European Union) is one of the reasons why most urban areas are now unsafe after dark. Drunken hooligans destroy property and threaten innocent passers by. Hospital emergency departments are full of people who have been injured deliberately rather than by accident. Departments which used to be called Accident and Emergency should be renamed Thuggery and Emergency.

It is any wonder that no one seems to have any respect for anyone else? (Let alone for authority). Can there be any surprise that many people now suspect that their government has deliberately allowed chaos to develop so that it will have an excuse to embellish and strengthen the fascist state with new laws?

In Britain, the state now provides everything for over a quarter of the population. There are millions of people who grew up in a household where the only source of finance was the state.

The End of the Interlude of Oil will mean the end for bloated, selfish societies.

Our modern civilisation is built on oil.

Oil gave us cheap energy. And cheap energy gave us our wealth, our progress and the complexity of the world we live in.

The absence of oil will take away our wealth and the complexity.

If, even when the oil is clearly running out, our leaders continue to take no practical steps to prepare for the end of oil then our world will become even more uncomfortable than I have described.

Instead of drifting back to a 19th century life, where fields are ploughed by horses and local communities make their own decisions, we could find ourselves living in a real nightmare: a complete collapse of society.

If this happens, gangs will patrol our cities mugging, killing and stealing. Disease will be widespread. Rotting bodies will litter the streets. We will enter the New Dark Age. There won't be any more oil, any more free energy, for another 500 million years.

Just as previous civilisations have completely collapsed, throwing their world into barbarism, so the same will happen to us.   And   our children will live in their history.

Chapter Five: A New Energy Blueprint

Threatened societies can survive, of course. Japan, in the 17th century, avoided collapse by limiting its use of natural resources. But, for that to happen we desperately need energy reform; we need a new energy blueprint.

'What can be predicted, with absolute certainty, is that the decline is coming, and our oil-consuming world is grossly unprepared for it. Somebody needs to get busy writing the script for Act II,' wrote Matthew Simmons in his book Twilight in the Desert.

We can't 'magic' more oil out of the ground. But there are several crucial ways in which our elected leaders can protect us from the coming disaster. We can't avoid the peak oil crisis. But we can prepare.

Politicians could encourage conservation, they could make serious (instead of dilettante) attempts to develop new sources of energy and they could change the way our society is structured. They could, and should, lead by example. That's what leaders should do. That's why we have leaders.

In a properly managed world our leaders would make an effort to find out exactly how much oil is left so that we can make best use of our resources - and so that we do not suddenly find ourselves looking at the last drop and wondering what happened.

We need to find out how much it is going to cost to extract the oil that does remain. At what price will it be viable to extract that oil? We need to be prepared for a world in which oil is going to become increasingly expensive. A barrel of oil will never again be sold for a few dollars. Barrels of oil will soon cost $100, $200 or more.

We need to decide on our priorities - which should not automatically include the military. How can we best make use of our diminishing oil supplies? Whose needs are greatest? There will have to be priorities and, to avoid profit taking arbitrageurs, individual countries will need to cooperate. Do the requirements of airlines and racing cars come above or below the needs of hospital generators?

No Government has yet made an effort to find the answers to any of these questions.

When it comes (as it will) the oil crisis will be made worse by incompetence, lack of vision, dishonesty, bias, commercial inspired prejudices, and a lack of leadership and direction.

`Energy conservation may be a sign of personal virtue, but it is not a sufficient basis for a sound, comprehensive energy policy,' said Dick Cheney, Vice President of the United States.

Your government probably hasn't told you this, but the International Energy Agency has already published the draft version of a report telling countries to prepare contingency plans to use when the oil shortfalls start.

The policies recommended by the report include:

1. Encouraging people to work from home and to use telephone, fax and computer.

2. A 50% reduction in the cost of public transport.

3. Reducing speed limits. (Though speed limits should not be reduced too much. Petrol consumption is increased at very high and at low speeds.)

4. Building more carpool lanes and making existing ones permanent.

5. Introducing driving bans on alternate days. (So that if your license plate ends with an even number you can drive on Tuesdays, Thursdays and Saturdays while if it ends with an odd number you can drive on Mondays, Wednesdays and Fridays.)

`We must face the prospect of changing our basic ways of living. This change will either be made on our own initiative in a planned way, or forced on us with chaos and suffering by the inexorable laws of nature,' said US President Jimmy Carter in 1976.

Our leaders should make our society less complex. They should encourage us to strengthen our family and community bonds so that we can become more locally dependent. In a world without oil we will not be able to buy fruit imported from half way across the world. We will need to buy food grown locally. Small individual communities will need to become more self sufficient. In a way this is what happened when the Soviet Union broke up in the early 1990s. Unable to continue competing in the arms race with the USA, and broken economically by the way the Americans had manipulated the oil price, the Soviet Union split itself into its former states. This helped by reducing the cost of government and administration. The change has been uncomfortable but it might have been much worse if the USSR had imploded and collapsed entirely.

Our leaders can easily reduce the amount of fuel we use moving ourselves around the country. They could improve public transport so that we will still be able to move about without our motor cars. We need a good, reliable national railway network. Countries which have reliable, public transport - buses, subway trains, trams and trains - will survive with much less trauma than countries which don't. Mass transportation systems need to be cheaper and more efficient. Train services need to be improved and prices should be subsidised. All internal flights should be banned. It is, for example, absurd that anyone should fly within Britain. The island is not big enough for people to need to fly from one city to another. It is, for example, absurd that travellers actually choose to fly from London to Birmingham. There are 37 flights a day from London to Manchester and hundreds more from London to Leeds, Newcastle and Paris. All these flights are unnecessary because there are excellent rail links available. The British Government has allowed air travel to grow at a rate of 13% a year and has allowed airports to expand. More runways will mean more flights. Improving (even subsidising) the railways would help prevent the use of oil (and protect us against global warming). So far the

British Government has deliberately created dependence on unsustainable, irresponsible, short haul flights by doing nothing to improve other forms of transport. Moreover, politicians have done nothing to cut down their own flying and, because they are worried about the short term political unpopularity, they refuse to discourage flying. Pop singers who fly around the world preaching about global warming are another irritant.

The biggest waste of oil is through traffic congestion. When thousands and thousands of motorists sit in their cars, inching forwards at a few miles an hour, the amount of oil that is consumed is phenomenal. Much oil is wasted not by drivers going too quickly but by drivers going too slowly.

In China roads which need repair are worked on at night, under floodlights, to cut down the number of traffic jams. The Government realises that the cost of paying overtime and setting up lights is far less than the cost to the economy of thousands of motorists sitting in traffic jams burning up petrol.

In other countries, however, motorways with lanes closed and no sign of activity have become a common sight. These unnecessary closures cost millions in terms of fuel wasted and hours lost. Arbitrary speed limits, often introduced on motorways solely as a way for governments and police forces to make money from speed cameras rather than in a serious attempt to reduce accidents, must be abandoned. Unnecessary traffic queues are also caused when lazy or incompetent officers fail to remove restriction signs after accidents and genuine repair works.

The following strategies would also help save energy:

Governments should introduce clear tax incentives encouraging people to work from home as often as they can.

Millions of gallons of fuel could be saved if governments subsidised school bus fares (and provided a better door-to-door service) so that parents no longer needed to waste time and energy transporting their children to school by car.

Every effort should be made to ensure that rural shops and post offices are kept open. Because of closures caused by predatory pricing by huge supermarkets, and by massive tax increases, millions of people now have to drive miles to buy a stamp or a loaf of bread. Village shops are the backbone of social life in rural areas. Closures will inevitably result in a massive waste of energy, an increase in congestion, an increase in pollution, and an increase in global warming. Governments should offer village and suburban shops preferential tax treatment in order to help their survival.

Governments must take steps to reduce their populations. Most countries are becoming larger and more complex. They consume more energy every year. Societies have, in the past, controlled their population growth successfully. Islanders on the tiny South Pacific island of Tikopia lived in near isolation for three thousand years. Their future was threatened by the fact that though they could only grow so much food on their small island their population kept growing. They dealt with this problem in two practical ways. They devised systems which maximised food production in a sustainable way and they introduced population control methods which kept the island's total population at around 1,300 people. A bigger example is provided by Japan. In 1650 Japan was in dire straits. Deforestation was causing soil erosion and lower crop yields. But Japan introduced an effective forestry management system and controlled its population.

Since energy is (and always has been) the driving force in our society governments should introduce fuel rationing. Undoubtedly unpopular, this will help delay the moment when the oil runs out. The world must do more to develop new energy supplies. Britain did this in the 16th century - when the nation (and the world) had its first energy crisis. In the 16th century the main energy source was wood. People used it for building houses, for building ships and for heating. Unfortunately, the forests started to run out and between 1500 and 1650 the price of wood rose eightfold. Britain had to rely on imported wood. The nation was saved by coal - a new energy source. From 1550 onwards the British used coal as a heating source in homes and workshops. Using coal enabled them to create new manufacturing processes. And by 1700 Britain had become the most productive economy in the world, ready to give the world the Industrial Revolution.

Governments must now make a real effort to develop alternative sources of energy to replace the disappearing oil. Alternatives won't replace oil (or the other fossil fuels) but they might help the withdrawal period less painful.

`When trouble is sensed well in advance it can easily be remedied,' wrote Niccolo Machiavelli. `If you wait for it to show itself any medicine will be too late because the disease will have become incurable. As the doctors say of a wasting disease, to start with it is easy to cure but difficult to diagnose; after a time, unless it has been diagnosed and treated at the outset, it becomes easy to diagnose but difficult to cure.'

Some people argue that we should make more use of the fossil fuel of which we have the most: coal. Coal is the dirtiest of all the fossil fuels. It's environmental enemy No 1.Electricity power stations fuelled by coal are the world's prime source of carbon emissions and, therefore, a major cause of global warming. In three years of operation a 1,500 MW coal plant will pump out three million tons of carbon dioxide. The biggest single producer of carbon dioxide gas in Western Europe is the Drax Power Station in North Yorkshire in Britain. The Drax Power Station burns coal and produces 21 million tons of carbon dioxide every year. That is the best part of a ton of carbon dioxide every second. And the power station works every second of every minute of every hour of every day of the year.

People who worry a lot about global warming want Drax to be shut down. If you don't think about things too deeply the protestors and campaigners have got a point. Shutting Drax down would reduce Britain's production of carbon dioxide by almost 4%. But it would reduce Britain's total electricity output by 7%. And that would mean that vast amounts of Britain would have to manage without electricity. In practice, Government policies have ensured that we will remain dependent on coal fired power stations such as Drax for a decade or more to come. It does, however, seem unfair that the coal industry pays nothing towards protecting the planet. Electricity generated by burning fossil fuels accounts for one third of entire human contribution to greenhouse gasses worldwide. There should be extra taxes on the use of coal.

The environment problems are not the only disadvantages with using coal. Coal is messy to dig out of the ground and the health care costs of digging it up are phenomenal. Coal kills far more people than nuclear power and it has been estimated that the health care costs of digging the stuff out of the ground are at least £50 billion a year. Another problem is that coal, like oil, is running out. It won't run out just yet but `peak coal' won't be reached all that much later than `peak oil' or `peak gas'. Another problem is that digging coal out of the ground and moving it around take a lot of energy. Most of the easily accessible surface coal has already taken. In the future coal will have to be dug out of deep mines. This will be expensive in terms of effort, money and human life. And the energy benefits (obtained by comparing the energy expended in digging out the coal with the energy obtained by burning it) will be modest.

Finally, coal's usefulness is limited. Steam trains may well come back as railway companies realise that using coal as a fuel is a real alternative. But we aren't likely to see coal powered cars or coal powered aeroplanes.

Coal may help us produce some electricity for a few years yet. But it isn't going to provide a long term solution to the disappearing oil.

Natural gas burns cleanly - though it still pollutes the atmosphere with carbon dioxide. Motor vehicles can be converted to run on it and it can be used to create the nitrogen fertilisers much loved by industrial farmers.

Natural gas fields in Middle East, Siberia and Alaska should last a little longer than the oil reserves.

But natural gas has to be liquefied and it is difficult and expensive to extract and transport. Moreover, natural gas is, like oil and coal, a diminishing natural resource; a disappearing fossil fuel. The number of drilling rigs looking for natural gas deposits is soaring but production remains virtually flat. When fields are discovered they tend to be small and used up very quickly.

As the oil runs out and gas is in greater demand so the cost will rise. Liquefied natural gas may provide a small answer to the problem of what to do when the oil runs out but natural gas won't replace oil. And, most importantly of all, it too is running out. If we convert vehicles to run on natural gas there will be less of the stuff left for heating. And the problems which we face will be delayed by a very small period of time.

Those who recommend that European countries should make more use of natural gas should remember that in the winter of 2005 Russia switched off the gas pipeline to the Ukraine because the pro-western government there refused to pay the extortionate prices Putin and the Russian Government demanded. The Russians just cut off the gas pipeline. It was mid winter. Temperatures outside were minus 32 degrees in some places. Temperatures inside were soon close to that. Children and old people died in freezing conditions. Belarus, Georgia and Poland have all also had their gas supplies cut by Russia.

That's the same pipeline that brings gas to much of Europe.

Half of the gas imported by the European Union countries comes from Russia. New EU members such as Hungary and the Czech Republic are virtually entirely dependent on Russian gas. Russia has power over Europe through its provision of gas supplies. But Russia is itself short of gas and is buying Central Asian gas to supply its customers. This means, of course, that the EU's supply of natural gas depends on whoever controls the pipelines. And that, of course, is Russia. Control of the pipeline inherited from Soviet Union also gives Russia control over the gas imported from Central Asia.

Politicians in Europe seem perfectly happy about this - and happy to separate their dependence on Russia for energy from their increasingly distant position politically.

However, rows of one sort or another seem to break out almost daily between Russia and the rest of Europe and Russia has made it perfectly clear that it is willing to use energy as a political weapon. In other words, if European countries don't do what Russia wants then Russia will cut off the gas supplies. Never, in the long history of the cold war, have European countries been as much at the mercy of the Soviet Union. Britain seems the most vulnerable of all. Decommissioning coal and nuclear power plants will mean that Britain's generating capacity will be cut by a third by 2015. Without oil as an option Britain will have to rely on Russia.

President Putin has already tightened his grip on the energy industry in his country. Russia has banned foreign owned companies from bidding to develop large oil fiends in Russia because the deposits are deemed to be strategic. And Russia has made it clear that it regards business relationships with foreign companies as acceptable only when they are acceptable. And always available for renegotiation.

Russia has bullied Royal Dutch Shell into ceding control of the Sakhalin-2 gas field project in far east of Russia to Gazprom, the Russian state energy giant, and has blocked BP's plan to develop a gas field in Eastern Siberia and kept foreign companies out of giant Shtokman field in Barents Sea. Russia has developed a magnificent way of dealing with western firms. It allows them to find oil and build the infrastructure and then it simply takes what as much of the oil and the infrastructure as it deems appropriate. Gazprom, the Russian company, has 60% of Russia's gas reserves and 17% of the world's proven gas reserves.

Gazprom has been buying up Europe's gas infrastructure and buying into electricity, oil and liquefied natural gas projects. Gazprom wants to be the biggest energy company in the world `Its not enough for us to meet 25% of global gas competition. We want to be the biggest energy company in the world,' said a Gazprom spokesman.

One British magazine found this rather ungentlemanly. Under the heading The Oil Thieves they wrote: `Russia and China are pursuing overtly nationalist policies in their grabbing of oil and mineral assets. And...Western firms - and shareholders - are losing out.'

But the Russians are not, of course, doing anything that the Americans haven't been doing for the best part of a century.

(Russia isn't the only country which is taking back control of its remaining oil. The President of Venezuela has taken back control of all the formerly private oil companies in his country.)

Wind is a clean, long established and renewable technology and the energy it produces doesn't require sending men underground to do dangerous jobs in difficult circumstances. There is no pollution (unless you count the alarmingly annoying noise pollution made by windmills). The energy produced by wind is relatively cheap and competitive with other sources of energy.

Unfortunately, you do need rather a lot of windmills to provide even a relatively small amount of electricity.

In his book `The Coming Economic Collapse' Stephen Leeb reports that in order to replace 10% of America's coal consumption (around 5% of America's electricity supply) with wind, would require between 36,000 and 40,000 windmills. To meet all America's electricity needs would require twenty times that number and take a decade to construct.

Inevitably, there are snags with obtaining energy from wind.

First, we would need vast numbers of windmills. There is no sign of anyone wanting to build them in sufficient quantities.

Second, building windmills uses up large amounts of fossil fuel. If we're going to build as many windmills as we need then we need to build them quickly before the oil runs out. But would the world's motorists, airline travellers and military agree to much of the remaining oil being used to make windmills? Somehow, I doubt it.

Third, to build enough wind turbines to fill our fuel needs would cost billions. One estimate suggests that building enough windmills in America would cost £2,500 per man, woman and child in America. Will citizens be prepared to spend £2,500 each today to provide themselves with fuel tomorrow? And will they put up with the environmental cost? Who will want the windmills in their backyard?

Fourth, most countries are not windy enough for windmills to work effectively. For example, even the UK is not as windy as some people think and the idea that wind turbines could generate even a fifth of the UK's energy needs by 2020 is nonsense. There are, quite simply, too many calm days when the windmills would stand idle. When the wind doesn't blow you don't get any electricity. The problems with wind reliability are vast. Germany has a lot of wind farms but these can generate only produce a sixth of their potential capacity. There is no way to store wind generated electricity when the wind blows very hard and there is no way that any developed country could obtain its power from wind farms unless the demand for power was dramatically reduced (to about a tenth of current levels.)

Fifth, wind turbines affect the environment too. They don't just look horrid and make an awful noise.(Try building one near to a country cottage owned by a politician to find out how unpopular wind farms really are.) If you erect enough to windmills to produce useful amounts of electricity they will change the climate and change the surface drag of the earth.

It is clear that even politicians don't have much faith in wind energy. In some parts of Europe, home owners who spend money on eco friendly wind turbines (or, indeed, on solar panels or other energy saving measures) will be liable to pay higher taxes.

The most practical use of wind and windmills may be to grind corn in small communities.

Solar power has been around a long time. The ancient Chinese used glass and mirrors to harness the sun's rays and make fire.

Solar energy sounds wonderful. It's clean and quiet and sunshine is abundant in some parts of the world. But solar cells aren't very efficient (converting only about a tenth of the light they receive into electricity) and they are expensive. Solar power is, at best, a niche energy source. It cannot be used to provide chemicals for farmers and it cannot easily be used to power the world's fleet of cars, lorries and aeroplanes.

It is perfectly possible to obtain electricity from moving water. Mill owners have been using rivers to power their water mills for centuries. Today, almost 10% of American electricity is obtained through hydroelectric projects.

Rivers and reservoirs aren't the only source of water power. The sea is another relatively untapped source and some people claim that we can harness wave power. The first problem is that waves vary constantly. Sometimes they are high. Sometimes they are almost invisible. The result is that the electricity obtained from the sea tends to be intermittent. And you can't run cars or jets on wave power. Nor can you make fertiliser from the sea. The second problem is that the energy required to obtain energy from the sea is enormous. Building and maintaining the infrastructure would require a tremendous amount of oil.

Hydrogen is a favourite among those who claim will be able to replace oil with something else. And, superficially, it looks good. Hydrogen fuel cells provide a renewable, clean energy source. They combine hydrogen and oxygen chemically to produce electricity, water and heat. But the main problem is that there isn't enough hydrogen to go around. In addition, hydrogen is neither cheap nor environmentally friendly. It costs a hydrogen fuel cell 100 times as much as an internal combustion energy to produce the same amount of energy. So, if you run your car on hydrogen fuel cells it will cost 100 times as much as it costs to run your car on petrol or diesel.

And although it's true that the by-product of using hydrogen as a source of energy is clean water (so the process is environmentally friendly), the other big problem is that collecting the hydrogen requires vast amounts of oil, natural gas or coal. There are no wonderful underground reservoirs of hydrogen waiting to be exploited. Hydrogen has to be manufactured from hydrocarbon sources such as coal or natural gas or extracted from water. So we're back where we started - needing a fossil fuel supply.

It is possible to generate small amounts of hydrogen using electricity generated by windmills and hydrogen fuel cells can be used to power cars so it might be possible to provide fuel for a few cars in this way. We would, however, need to build an entirely new infrastructure to support and fuel hydrogen powered cars.

Finally, the big, insurmountable problem is that producing hydrogen always uses more energy than is obtained. Hydrogen, like biofuels, is a negative source of energy. And it just doesn't make sense to turn oil, or anything else, into hydrogen.

Some `greens' and self-styled `environmentalists' believe that it is possible to replace oil and to control the amount of carbon in the atmosphere by using solar and wind power and being more efficient and careful in our use of coal and gas. But to reduce the use of gas and coal by, for example, putting up prices enough to force energy conservation, will mean massive unemployment and mass starvation. These are no worse than the problems we are going to have to face when the oil runs out but these are not things that politicians will ever do voluntarily.

A growing number of thinking greens now recognise that nuclear power really has to incorporated into our energy sources. James Lovelock, is probably the world's best known and most respected environmental scientist. He is the inventor of the Gaia theory (that the earth behaves like a living organism and actively sustains its climate and chemistry to keep itself habitable). He argues that he believes the Earth to have reached a dangerous condition. `Green lobbies,' he says, `are well-intentioned, but they understand people better than they do the Earth. Consequently, they recommend inappropriate remedies and action. Wind turbines and bio-fuels alone will not cure the Earth's sickness.'

Lovelock recommends that nuclear energy, as part of a portfolio of energy sources, would make good medicine for the Earth's ills. He points out that by the time Greenland's icy mountains have melted the sea will have risen seven metres, making low lying cities such as London, New

York, Tokyo, Calcutta and Venice uninhabitable. A four degree rise in temperature will eliminate the vast Amazon forests which are a great global air conditioner. Extra heat from greenhouse gases, the disappearance of arctic ice, the changing structure of the ocean's surface and the destruction of tropical forests will be amplified.

Lovelock says that green concepts of sustainable development and renewable energy are beguiling dreams that can lead only to failure. `I cannot see the USA or the emerging economies of China and India cutting back,' he says. `There is no sensible alternative to nuclear energy. We need something much more effective than the green ideology of the Kyoto agreement.'

The simple unavoidable truth is that wind and solar energy are temporary sources of energy. They work when you have wind or sun. So, only those people who are happy to read, use their computer or watch television only when the wind is blowing or the sun is shining will be happy to rely on wind and solar energy for their electricity. Nuclear power provides a reliable energy source.

Other countries have for some time been increasing the amount of electricity they obtain from nuclear power. In France 79% of electricity is now generated by nuclear plants. In Japan its 30%. In Germany its 31%.

China's leaders are well aware that oil is running out fast and so China is now working hard to acquire enough uranium to run the thousands of nuclear plants it knows it will have to build (and has already started building). China, where the uranium demand will rise by up to 6 times by 2020, currently has nine one million KW nuclear power generators and is planning to build three of the same size each year for the next 15 years. The Chinese have also said that they will build strategic reserves of uranium. In 2006 China obtained less than 2% of its energy needs from nuclear plants. By building three large nuclear power generators a year they will double this percentage.

That means, of course, that they will still have to find the vast majority of their energy supplies from coal, oil and other sources.

Even the Arabs are keen to use nuclear energy, though the Americans are opposed to their building nuclear power stations. The Arabs say that nuclear power is the energy of the future and (not unreasonably) that no one has the right to stop them using it. They recognise that their oil and gas supplies are fast running out and they want to sell what they've got left, rather than use it up themselves. Iran, still one of the world's main sources of oil, is one of the countries which wants to convert to nuclear power internally and to sell the oil it produces to outside countries. This, they say, will enable them to continue making money and to have the cleanest fuel themselves. No wonder the American Government cannot understand: it's a policy that makes good sense.

(If the Arabs, who hold most of the world's remaining oil, want to use nuclear power what does that tell us about the remaining oil stocks?)

Britain, in contrast, has been woefully slow to build nuclear power stations. According to one study demand for energy in Britain could outstrip supply by 23% at peak times by 2015 - even without a global oil shortage. As ageing coal and nuclear power stations have to be closed down Britain's generating power will be cut by a third. Declining North Sea oil and gas production will make things even worse. Britain will become increasingly reliant on imports of oil and gas from countries which don't much like the British Government and which have ready markets for their products elsewhere.

The British Government has produced a white paper confirming its commitment to nuclear energy but, even though time is running out rather rapidly, they have neither built any nuclear power stations nor encouraged (or even allowed) any commercial companies to do so. The white paper does, however, state that when the Government finally makes up its mind to take the plunge and allow nuclear power stations to be built: `it would be for the private sector to fund, develop and build new nuclear power stations'. The private sector will have to meet the full costs of decommissioning any power stations and the cost of dealing with the waste. This means that only very, very large companies will be able to consider such a risk. And it means that even though the British Government may agree that without nuclear power the lights will go out it is going to rely entirely on private companies to build and run the power stations it says the country needs. It is not easy to see just why Britain has a Government at all when the nation's energy future would be just as secure without one. It takes years to build a nuclear power station. Even if the British Government authorised the building of nuclear power stations today it would be ten years before they were operational. Britain is, without a doubt, one of the countries least well prepared for peak oil. The 12 nuclear power plants in existence in Britain currently provide 20% of the nation's electricity. Sadly, they are all old and beginning to break down. They have already outlived their 20-25 year life expectancy. Much of Britain's energy supply involves oil - the stuff that is running out. Now that Britain's own North Sea oil supplies are running out the country is a net importer of oil and gas and is totally dependent on its ability to import oil from other countries and, therefore, on the goodwill of countries such as Iran, Nigeria, Russia and Venezuela.

Nuclear power is clean, effective and relatively safe. France, which gets most of its electricity from nuclear power, has the cleanest air in the industrialised world and the cheapest electricity in Europe. The French do not store their nuclear waste. Instead they reprocess it. Instead of burying spent fuel rods deep in the sea or underground they have built a massive plant on the coast of Normandy to recycle the used fuel and so reuse it.

Those who complain that nuclear power isn't safe should know that every year the deaths caused by coal mining exceed the number of deaths associated with the entire history of nuclear reactors.   In   April 2007 there were 103 nuclear plants operating in the USA. These have produced 20% of the nation's electricity without any major incident since the problem at Three Mile Island. The infamous Three Mile Island accident killed no one.

Lighting a candle is dangerous. Having a bonfire is dangerous. But if you're measuring safety then nuclear power is to coal mining what passenger flight is to bungee jumping. In the last 28 years there have not been any serious incidents at any of the nuclear plants operating in the USA (plants which provide 18% of America's electricity). Since the nuclear submarine Nautilus was launched in 1954 American nuclear ships have travelled more than 150 million miles and gone round the globe 40,000 times without a single nuclear incident. The WHO estimates that 50 people died as direct result of the infamous Chernobyl disaster, which involved an obsolete, badly maintained reactor.   It is estimated that another 4,000 deaths can be blamed on the 1986 accident. Since that time around 200,000 coal miners have died as a result of coal mining. (And the WHO says that 150,000 deaths a year (and the figure is rising fast) are a result of global warming.)

Both China and South Africa are building advanced power plants - to protect themselves from rising coal and natural gas prices and to meet new restrictions on carbon dioxide emissions - and the plants they are building seem extraordinarily safe. During a safety test at a Chinese reactor engineers did their best to create a diaster. They cut off the flow of the coolant that removes heat from the nuclear reactor and then withdrew the control rods - usually a recipe for meltdown. The reactor simply shut down with no damage or threat.

Nuclear power produces virtually no carbon dioxide and is very climate friendly; it is, it seems, the only cost effective and environmentally acceptable way of creating electricity.

Nuclear power doesn't provide all the answers. It certainly doesn't provide an alternative fuel for motor vehicles, aeroplanes and ships. But it's a start and those countries which have nuclear power stations will at least be able to provide their citizens with heat and light.

Those who oppose nuclear power point out that uranium is a finite resource. This is obviously true. However, it is usually fairly easy to mine, and can be reused, and there is much dispute about how much uranium there is left. Some experts say there is enough to last for centuries. Others, more pessimistic, suggest that the supply of uranium will completely run out within twenty years. Who do you believe? I have no idea. There are so many hidden agendas, lobbyists and lying politicians around that it is quite impossible to know the truth.

At the moment America relies heavily on Russia for half its uranium. The USA has a deal with the Russians whereby 20,000 Russian nuclear weapons are being converted into fuel for American nuclear reactors. But the deal ends in 2013 and China, India and Russia are all building nuclear reactors as quickly as they can. At the moment uranium production worldwide meets about 65% of reactor requirements. That's a considerable shortfall. Most worryingly, there has been little investment in new uranium mines for 20 years because the price has been so low and nuclear energy so unpopular. Recent rises in uranium prices have boosted exploration and mining activity. But it takes time to find uranium and to start digging it out of the ground.

Breeder reactors reprocess spent fuel but these are extremely expensive to build.

It has been estimated that the 10,000 nuclear reactors America needs to build would use up all the known uranium supplies within the next decade or two. And then the Americans would be left with 10,000 very expensive but entirely useless nuclear reactors.   The only hope is that miners will be able to find more uranium. There are a number of hopeful sites around the world.

The waste from nuclear power stations is tricky to deal with. The direct waste is around 1,000 metric tons per plant per year but uranium tailings (residues from the mining process) are also radioactive and can be as much as 100,000 metric tons per nuclear power plant per year. These are, however, problems we have to deal with unless we want to live without electricity.   Scientists are working to find better ways to get rid of it. For example, the waste can be vitrified - converted into a glass like solid - and research is continuing into making synthetic rock, with immobilised radio active elements. Another possible solution is to remix the high level waste with lower level waste to create a product that is below the original radioactivity of the uranium dug out of the ground. This can then be dumped into the empty uranium mines where the radiation produced by the stuff placed in the hole will be no more than the radiation produced by the original uranium ore before it was dug out.

Critics also point out nuclear reactors use a lot of water. They do indeed. But the water that has been used isn't `used', it's just borrowed to cool the reactor and then recycled. It is warmed. Clever scientists could, no doubt, find a way to harness the heat from the water.

And critics say that nuclear reactors are expensive to build. Again, this is true. But the cost of building them can be reclaimed by selling the electricity they produce. So, what's the problem?

I used to be oppose nuclear power.

But unless we are all prepared to go to bed when it gets dark and to stay in bed when the weather gets cold there really isn't another sensible option. How many of those who oppose nuclear power will be happy to turn off their television sets, radios, computers and dish washers?

Or maybe those who oppose nuclear power prefer biofuels and are prepared to put up with mass starvation in poorer nations so that they can continue to use their computers?.

Nuclear power is inevitable. Gas prices are going through the roof, oil is running out and everyone is scared stiff of greenhouse gas emissions.

Governments around the world should provide tax incentives to encourage people and companies to make and use products which last longer and can be repaired. This would enable us all to use our limited and disappearing resources more wisely.

Our just-out-of-warranty video recorder broke the other day. The cheapest call out fee for a man to repair it was £50. On top of that there would, of course, be a fee for any work required and additional costs for replacement parts. A brand new, middle of the range replacement cost £59.99. So the sensible option was to toss a large piece of electrical equipment out with the garbage and to buy a replacement.

If governments were serious about recycling, and protecting the environment and the world's disappearing resources, they would subsidise the repair of electrical equipment by providing grants and tax breaks to encourage more repair workers. If governments really wanted to help global warming they would slap a huge tax on tumble driers and hand out free clothes lines and clothes pegs. (The clothes line is the best and most efficient solar and wind gadget known to man.)

Governments should also be making a real effort to encourage small, local manufacturing companies. When the oil runs out we won't be able to import shoes, trousers, skirts and bras from China. Communities which don't make these items themselves will have to do without them.

We should prepare children and teenagers for the future we will bequeath them. It will be nothing like the future they might have imagined.

We are all constantly told that we must learn to adapt to change. The young take great delight in taunting their elders as previous generations struggle to cope with video recorders, DVD players, digital cameras, computers, fax machines, e-mail and so on.

`Change,' say the techno-addicts, `is here to stay. You just have to get used to it.'

But the truth is that the young don't know what change really means.

To them `change' is a new variety of Play Station or a new search engine.

They're going to have a big surprise because the `change' that is coming will be like nothing they've ever known or imagined.

They will find themselves living in a science fiction fantasy land - the sort of world depicted as `life after nuclear war'.

The best thing we can do the next generation is to encourage them to acquire simple, practical skills.

The kings of the future will not be bond traders or software engineers. The kings of the future will be carpenters and farmers. There will be no need for airline pilots or industrial chemists. Road menders and bureaucrats will be unemployable. Anyone who can mend a wood burning stove or grow potatoes in a small patch of garden will be much in demand.

Local farmers must be encouraged. And if hundreds of millions are not going to starve to death, the world has to turn vegetarian.

The world's population is growing constantly - particularly in Asia. As the world gets wealthier (particularly in China and India) more and more people will start eating meat (because they want to do what is considered fashionable in the West).

And so much of the world's supply of grains will be used to fatten up animals.

But, at the same time, the global shortage of oil will also mean that a good deal of corn will be used to create ethanol as an oil substitute.

This will mean that the cost of grains will go up.

In addition, of course, global warming will mean that growing crops will become more difficult. This will also add to the price of corn.

The surge towards meat eating will exacerbate the world's problems. Most of the world's population will eventually be forced to become vegetarian.

Growing grains to feed to animals to eat the animals is very inefficient.

Here are some facts you should know:

a) One hundred acres of land will produce enough beef for 20 people but enough wheat to feed 240 people.

b) If we ate the plants we grow -instead of feeding them to animals - the world's food shortage would disappear virtually overnight.

c) Half the rainforests in the world have already been destroyed to clear ground to graze cattle to make beef burgers. (The remaining rainforests are now being destroyed to make room for growing biofuel crops.) The burning of the forests contributes 20% of all greenhouse gases. Roughly 1,000 species a year become extinct because of the destruction of the rain forests. Approximately 260 million acres of US forest have been cleared to grow crops to feed cattle so that people can eat meat.

d) Every year 440 million tons of grain are fed to livestock - so that the world's rich can eat meat. At the same time 500 million people in poor countries are starving to death. Every six seconds someone in the world starves to death because people in the west are eating meat.

e) Approximately 60 million people a year die of starvation. All those lives could be saved (because those people could eat the grain used to fatten cattle and other farm animals) if Americans ate 10% less meat.

f) African countries - where millions are starving to death - frequently export grain to the developed world so that animals can be fattened for the dining tables of the affluent nations. If all those people who support concerts to raise money to draw attention to the starving millions in Africa chose instead to turn vegetarian there would be no starving millions.

g) The world's fresh water shortage is being made worse by animal farming. Meat producers are the biggest polluters of water. It takes 2,500 gallons of water to produce one pound of meat. If the meat industry in America wasn't supported by the taxpayer paying a large proportion of its water costs then hamburger meat would cost $35 a pound.

h) Cows produce methane which also increases global warming.

To survive we will have to get rid of big farms (the American way of farming) and replace them with small farms (the French way of farming).

In America (and to a large extent in most other countries too) farming is done by employees sitting on large tractors. They receive a monthly pay cheque from the conglomerate which employs them. They have no time to understand what they are doing. And no incentive to care.

At the other extreme the smallholder who looks after a small farm of ten acres or so will know everything there is to know about his land. He will care. He will understand the importance of getting the best out of his land. He will be prepared to intercrop plants whose roots go to different depths. He will alternate crops to keep the soil in tip top condition and to reduce the risk of pests. He will encourage those species of wildlife which are necessary to the success of his farming. He will select crops which make the best use of the particular strengths of his land. He will know and understand the land. He will notice when things happen. And if he doesn't know what to do he will find out.

It is surprising, but small farms invariably produce far more food per acre than large farms. Whether you measure the output in terms of tons of food produced or money earned, large farms are inefficient when compared to small ones.

And, most importantly, small farms use all their vital resources far more effectively. They use land, water and oil more efficiently. They understand that animal manure is a free gift, far more useful than any bag of chemical fertiliser, not a health hazard which must be disposed of.

Agronomist Jules Pretty has, over the last decade, studied sustainable agriculture projects in 57 countries. He has found that sustainable agriculture increases food production by an astonishing 79% per acre. Pretty found that yields rose 73% for 4.5 million small grain farmers in Asia. A study of 14 projects, involving 146,000 farmers in the developing world, showed that using practises such as fighting pests with natural adversaries rather than chemicals increased production by 150%. When rice farmers in Indonesia got rid of pesticides completely their yields stayed the same but their costs fell dramatically.

In farming, small really is beautiful.

Our land has been turned into large farms for all the wrong reasons.

Big corporations can get access to money far more easily than small farmers. Big corporations can cope with the paperwork more readily (they can hire a team of bureaucrats to fill in the forms) and are, therefore, at a huge advantage when it comes to claiming grants and farm subsidies. Big corporations can deal with all the regulations more easily for the very good reason that in many cases the regulations were devised for, or on behalf of, the large corporations.

We have, in the past, allowed oil, large tractors and big corporations to take over farming. We have replaced people and passion with synthetic fertilisers and regulations.

If we continue to eat then we have to change.

Much of what I have suggested would result in long term zero economic growth. But my simple proposals would help allow our society to wind down and to become more sustainable and less wasteful. As a side effect, most people would probably be happier.

A simpler society does not mean an unhappy society.

There are snags.

First, our political leaders will not accept zero economic growth. Our politicians have created a political and economic environment which can only survive if there is growth. Without growth there will be no increase in the government's tax revenue. And without a steady and considerable increase in tax revenue governments will be unable to meet their financial commitments or pay their considerable debts.

And none of my proposals will work if one group of leaders take unilateral action. For my plan to work every leader in the world will have to work together, and agree to make the necessary changes.   And that won't happen.

The Americans, for example, have steadfastly refused to make even small changes that might help cut down their use of fossil fuels. They claim that to do so would damage American industry.   Despite being an oil importer the Americans still subsidise the cost of petrol (a gallon of petrol would cost five times as much in America if Americans paid the real price of the oil).

It's easy to put all the blame on the George W.Bush administration but the Clinton Administration, with Al Gore as vice president, was just as guilty. Clinton, too, subsidised oil so that Americans could buy cheap petrol. When Clinton and Gore took office in 1993 environmentalists hoped that their administration would continue the work of energy conservation and renewable energy programmes begun under President Jimmy Carter. But very little happened and few significant energy policy changes were made between 1993 and 2001. Coincidentally, Enron had, of course, made donations to Democrats as well as Republicans.

The bizarre American attitude towards oil and energy isn't confined to Presidents and Vice Presidents. Not long ago, Spencer Abraham, an American Secretary of Energy suggested that the energy crisis could be solved by removing regulations and building more pipelines and refineries so that Americans could consume more oil and gas.

Lists of potential global crises, devised by politicians or journalists, rarely even include energy or peak oil.

Governments don't want to talk about the coming oil crisis. That will be unpopular. They don't want to introduce taxes on oil use. That will be very unpopular. And it seems to me extremely unlikely that politicians will agree to do any of the things they need to do.

Is there any serious hope that Americans will stop using oil hungry cars and learn to survive without air conditioning? Is there a chance that China will decide to halt progress and stay where it is? Is there a chance that airlines will voluntarily ground their aircraft - or that governments will force them to?

Of course there isn't.

On the contrary, everything our leaders have done so far seems to have been designed to make things worse.

Preparing the planet for the coming crisis would require politicians with intelligence, honesty, creativity and initiative. We desperately need thoughtful, courageous, creative politicians to ask the right questions and find some good answers.

But we don't have thoughtful, courageous, responsible politicians. And our politicians certainly do not care for our welfare.

No politicians have publicly acknowledged the problems of peak oil. No politicians have discussed the questions that I've raised in this book (let alone tried to provide any answers).

Around the world I cannot see any major political leaders who understand the size of the problem and who might be prepared to try to force through the oil usage cuts which would be required to give us a sensible chance to `kick' our oil habit at a respectable rate.

Attempts have been made to make it easy for politicians to take the steps that are needed. Richard Heinberg's book `The Oil Depletion Protocol' describes a simple way to ease the pain. `The protocol itself is so simple,' he writes, `that its essence can be stated in a single sentence: signatory nations would agree to reduce their oil consumption gradually and uniformly according to a simple formula that works out to being a little less than three per cent per year.'

It is a good and noble proposal and one that deserves to be taken seriously. It could help us avoid much pain. It could help prevent the wars, the terrorism and the economic disasters that lay ahead.

But do you believe that America would willingly promise to cut its oil consumption by three per cent per year? Do you believe that China would abandon growth and accept a cut in its oil usage?

Sadly, nor do I.

I cannot think of any major country where the leaders have the necessary qualities. No government has made a real effort to prepare its people (either as a nation or individually) for the coming energy crisis; the greatest crisis our civilisation has ever faced.

On the contrary many countries (particularly America and Britain) have accumulated such enormous national debts that they cannot survive without economic growth.

And for real economic growth nations need to continue to use, and rely on, vast amounts of oil.

If you and I are to survive the coming disaster then we need to prepare ourselves, our families and our friends for the future. We cannot and should not rely on politicians taking the correct decisions for us.

The final two sections of this book contain practical advice which will, I hope, prove of help.

Chapter Six: Your Personal Survival Plan

Having read this book you now know far more than 99.99% of the population know about the future we face together.

The first question you will probably want to ask yourself is: is this really true? If, despite all the evidence I have quoted, you doubt the truth about peak oil (and suspect that the politicians are right and the experts have got it wrong) then you must look at the odds.

Is there a 50% chance that the politicians could be right? Or do you think the chances that the politicians are telling the truth might be as high as 90%? Do you think someone might, after all, invent a perpetual motion machine and provide us all with everlasting supplies of free energy?

Now, change the problem.

If you knew that there was a 10% chance that your home would burn down in the next five years would you be alarmed? Would you take whatever action you could to prepare yourself and your loved ones?

The second question, which follows on from the first, is what are you going to do with the information you've acquired?

There will, of course, be many people who will choose to ignore the truths in this book in the same way, I suppose, that many people deliberately ignore the truths about the relationship between smoking and cancer and eating meat and cancer. In a world where people buy cigarettes, light them and put them in their mouths when the packets in which the cigarettes are distributed contain large, clearly printed health warnings we should not, I suppose, be surprised by anything.

When faced with an unpleasant truth most people prefer to draw the curtains and turn on the television set, hoping that when they draw the curtains back the truth will have gone away.

But the truth isn't going to go away.

And one day, when the electricity goes off the people who don't want to face the truth will be forced to draw back the curtains.

And then they will have a very nasty surprise.

You, I hope, will be prepared.

This part of the book is designed to help you prepare yourself for the unavoidable crisis that is heading our way.

If you are not prepared, you will not survive.

For the first time in a long time, life will be about the survival of the fittest.

There are two things of which we can be certain.

First, that the Government will make no real effort to prepare the country, or us, for the coming chaos.

Second, when the problem arrives they will make a mess of handling it, limiting themselves to filling the streets with policemen and soldiers to bludgeon those involved in the inevitable angry protests.

So, you must make your own preparations.

Here is what you must do

1. Prepare yourself mentally for a different world. A world in which the rich ride horses, the middle classes use bicycles and the poor walk

everywhere they want to go. Think carefully about your current lifestyle. And try to imagine how difficult (and different) things will be when there is no oil.

2. Energy prices are going to rise inexorably. Take time now to reduce the amount of energy you use. Cut out all non-essential energy usage. Within the home the greatest expenditure is usually heating. See how low you can turn down your thermostat and still survive comfortably. Wear a sweater indoors and you may be able to cope with a lower temperature.

3. If possible you should acquire alternative forms of heating and cooking. Do not rely on one energy source. If you have gas central heating then you should have one or two electric heaters available. If you have to replace your oven consider purchasing one which will enable you to cook with either gas or electricity.

4. If you can become at least partly independent by installing an alternative personal energy source then now is the time to do it. Maybe a small windmill will supply at least part of your electricity needs. If you have a working but unused fireplace in your home then have the chimney swept and cleared so you can have log or coal fires to keep warm. Start laying down stocks of logs and coal. These things won't rot and I don't think there's much chance that they are going to go down in price.

5. Look around your home and make a list of all the gadgetry and equipment upon which you are dependent. How will cope without each item? Can you accumulate spares? Can you learn how to repair any of these items?

6. Prepare yourself for electricity blackouts by buying lamps and candles. Don't forget that you will need candle holders and matches. (And make sure that everyone in the family knows how to use them safely.)

7. If you are considering changing your motor car you might consider choosing a car which uses less fuel. Look also for a vehicle which has a

decent tank capacity so that you can continue to make small journeys when there are fuel shortages. Reconsider all your travel needs. How much do you need a car of your own? Would you be able to cope more economically (and with less hassle) if you simply relied on taxis and hire cars occasionally? How big a car do you really need? Must you buy a new car? An older car may need more maintenance but the maintenance will almost certainly be easier to manage than a car which is controlled by a series of complicated computers. If you don't have a bicycle this would be a good time to purchase one. If you can't ride one then now is the time to learn. Folding bicycles are easy to fit into a car and it should be possible to carry them onto public transport. Equip your bicycle with panniers and a basket so that you can carry shopping on it.

8. If you are choosing a new home consider your likely future needs. Houses within walking or cycling distance of a railway station will sell at a premium in the future. A home that has its own water supply will be particularly attractive as public water supplies come under threat. But look for a spring or gravity fed supply rather than a bore hole. In order to get water out of a bore hole you will need an electric pump - and when the electricity goes off you will get no water.

9. If you have land consider establishing a vegetable garden where you can grow at least some of your own food. Try to grow as much food as you can. If you have little or no experience of gardening it will probably take you a year or two to learn some basic gardening skills. Acquire a small library of relevant gardening books. Try to manage your garden without using artificial fertilisers, pesticides, herbicides or other chemicals. Even if you don't want to start your own vegetable garden straight away you should, perhaps, start thinking of living in a home where a vegetable patch would be possible. (And, ideally, you should have a vegetable patch which is not open to the world. Thieves of the future will not be stealing

television sets and mobile phones. They will be stealing potatoes and runner beans. If you grow your own vegetables you will have to be prepared to protect them from thieves. You should prepare now by making your house formidable, impenetrable and uninviting. If you need to dig up your front lawn in order to grow more food, you will need to think about ways to protect your crops from thieves?)

10. Do not take on any additional debt. Try to pay off any existing debts as soon as you can. Credit card debts are particularly expensive and can be a huge drain on your personal resources. If interest rates soar your repayments could be crippling. Water and food are, like fuel, going to become extremely expensive. And the coming price rises in oil and food will be structural not cyclical. Oil and food will never again be as plentiful or as cheap as they are now. Your savings could help you survive.

11. This could be a good time to examine your life. How many of the things you spend money on are essential to your health and happiness? How many of the things you buy turn out to be a burden rather than an asset? Every time you make a big purchase consider not just the cash price but also the time price. How many hours did you have to work to earn the money to pay for it? If you are contemplating buying an electrical item that costs £500 and you earn £5 an hour net of taxes then the item you're thinking of buying will cost 100 hours of your life. Step off the consumer treadmill and you may feel physical and mental benefits.

12. Try to replace some of the more complex tools in your house with simpler tools that don't need electricity. For example, a small hand drill may be slower and harder to use than an electric drill but you will still be able to use it when there is no electricity. Accumulate simple well-made hand tools to use around the house and garden.

13. This might be a time to start learning simple, practical skills so that you will be able to look after your home and your belongings without always

being reliant on outside `experts'. Learning basic carpentry and basic plumbing will provide you with considerable freedom.

14. In recent years it has become increasingly difficult to obtain the services of a general practitioner out of hours. This is likely to continue (if not to get worse). Hospitals are likely to deteriorate still further as they struggle to cope with a top heavy bureaucracy, an increasingly incompetent and unhappy workforce and an ongoing energy crisis. You should, therefore, make sure that you acquire some simple medical skills. Put together a simple first aid kit and a small library of easy to understand medical books.

15. Try to do as much shopping as you can at local stores and local markets. When buying food try to buy locally grown food. Big supermarkets may sometimes (but not always) be cheaper and it is certainly more convenient (if rather soul destroying) to do all your shopping in one store but when oil becomes increasingly expensive the big stores will not survive. (Transporting food and other supplies to their stores will be costly and many of their customers will no longer have the transport available for them to visit out of town stores.) If you and your neighbours do not keep small shops and markets alive where will you shop when the supermarkets close down?

16. Try to limit the amount of rubbish you accumulate. As oil become increasingly expensive and local councils struggle to cope with their dramatically increasing pension obligations so local services will deteriorate considerably. Rubbish collections, already threatened, will be non-existent. Try to free your home of as much rubbish as you can now. And be cautious about taking home new rubbish and clutter. You will need to find new ways to get rid of your rubbish in the future - either by burying it or burning it.

17. You need to prepare yourself to cope with the effects of global

warming. Learn to dress for the weather not for the season. Be prepared for sudden and apparently inexplicable changes in the weather. Prepare your home to cope in high winds and be prepared to survive through long periods of drought. If you live in an area which is likely to flood then think about permanently moving your most valuable possessions upstairs.

18. In the medium and long term our lives will be much more concerned with local issues than with national or international issues. Wherever you live it is important that your community should be designed for a world in which fossil fuels play a smaller and smaller part. This means that communities will inevitably become smaller and more self-contained. It must be possible for people to walk or cycle from home to work and from home to the shops. Communities must be built with pedestrians rather than motor vehicles in mind. Public transport must be improved dramatically. There is no point in spending vast amounts of money on huge road building programmes. Large administrative buildings are also a waste of money because within a relatively short period of time political and administrative control will become increasingly local. Local and national politicians must be persuaded to think of, and plan for, a future without oil.

Epilogue

When the oil runs out (which it will do soon) I believe that the world will change for ever.

Is there a chance that none of this will happen?

Yes, of course there is.

Someone may discover another entirely `free' form of energy: a form of energy we can use to drive motorcars and aeroplanes and from which we can obtain electricity.

Or maybe explorers will discover a huge oil field four miles underneath Milton Keynes.

Perhaps a young scientist in Latvia will perfect a perpetual motion machine.

Who knows, perhaps someone will find a way to turn sea water into oil.

Anything is possible.

If you think any of these scenarios are likely (and you're happy to put your trust in chance and good fortune) then you have no need to worry.

Otherwise, I think you should take this danger very seriously.

As I wrote at the beginning of this book, the problem of the disappearing oil is a threat to our civilisation much greater than global warming or terrorism. It is a threat which everyone in power knows about but everyone in power steadfastly ignores. It's a danger no one talks about.

Appendix

Here's a summary of the ways that oil has changed our lives.

1. Transport

a) Building cars takes up a great deal of oil. Running cars depends on oil. (The average American motorist consumes his or her body weight in crude oil each week.) And the roads which society has had to build so that motorists can use their cars are built with fossil fuel products. Asphalt contains vast quantities of oil and, of course, road building machinery runs on oil. The success of road transport did enormous damage to rail transport in many countries (particularly America and Britain). Roads are built by governments. Rail tracks have to be built and maintained by the railway companies so trains never stood a chance. `Globally,' writes Richard Heinberg, `cars outweigh humans four to one and consume about the same ratio more energy each day in the form of fuel than people do in food. A visitor from Mars might conclude that automobiles, not humans, are the dominant life form on planet Earth.'

b) Aeroplanes need oil. There are, as yet, no commercial aeroplanes flying which use coal, nuclear power, hydroelectricity, solar power or natural gas. Only gliders, hot air balloons and kites are exceptions. Americans are almost as dependent upon aeroplanes as they are upon cars. In an average year the average American (were there such a creature) would take an astonishing 2.85 trips by aeroplane.

2. In Agriculture

a) Simple, traditional agriculture works in a modestly energy efficient way. A man, helped perhaps by one or two animals, plants seeds. He then relies upon nature to water them, the soil to provide nutrients and the sun to provide energy. Later he harvests what he and God have grown together. There is a small energy gain. The free energy provided by the sun ensures that the man gets back more than he put in. At the start of the 20th Century two German chemists called Bosch and Haber invented a way to combine atmospheric nitrogen with hydrogen to make ammonia. Some argue that their discovery was the most important single invention of the 20th century. The synthesis of ammonia provides almost all the inorganic nitrogen used on farms. The amount of inorganic nitrogen produced is approximately equal to the natural nitrogen tonnage produced annually. That means, of course, that the Haber-Bosch process has doubled the available nitrogen on the planet. It's this that has enabled farmers dramatically to increase food production during the 20th century. Naturally, the extra food which has been produced has enabled the world's population to grow. Populations have frequently grown faster than local food production and so rich nations, such as the USA and the European countries, have used force and bribery to force underdeveloped nations to grow food for them. It was this that led to the introduction of the phrase `banana republic' to describe a country using all or most of its land to produce a single crop. Native farmers have been persuaded or forced to use artificial fertilisers, and even genetically engineered crops, in order to boost production and reduce prices. Except in the European Union (where the Common Agricultural Policy has subsidised and therefore sustained small farmers - particularly in France) the result has been the death of the small farm. Food, today, travels an average of 1,300 miles from farm to diner. That's a lot of energy being wasted just to transport food. (One downside to all this is that all that the extra nitrogen has produced massive pollution of streams and rivers. Another downside is that people in poor, developing countries often starve to death while the food their nation grows is taken to feed cattle so that rich Americans can eat hamburgers.)

b) Tractors, combine harvesters, trucks and lorries have altered the way farm produce is harvested and distributed. Before tractors were widespread around a quarter of all agricultural land was used to produce the feed needed for horses and bullocks pulling carts and ploughs. Tractors enable one man to cover vast areas of farmland and have, therefore, greatly increased farm production on large farms. Today, cheap fossil fuels ensure that by using petrol driven machinery the farmer can dramatically enhance the energy he puts into his farming. Farming has been turned into an industrial process, favouring large scale operations. (It's easier and more cost effective to use large pieces of machinery when you're farming vast, open areas.)

c) Petrochemical based herbicides and pesticides have been widely used for half a century and have altered farm yields. There have also been dramatic reductions in the safety of food produced.

All these changes have had massive effects on society (fewer people now work on the land) and on the environment (pollutants have done great damage). Power provided by fossil fuels has enabled us to refrigerate and transport food (and to refrigerate again when it is stored in our homes). These advantages will all disappear when the oil runs out.

3. Warfare

When the 20th century began wars were fought by foot soldiers and cavalry. Ships were powered by coal. Gradually the military have discovered the joys of oil power. Oil has enabled soldiers, sailors and airmen to kill far more people (especially civilians) with much greater efficiency. Oil is so important to the various branches of the military that during World War II the Allies tried to stop the Germans and the Allies both tried to stop one another obtaining oil supplies.

Without oil warfare would go back a century. There is no doubt that when the oil really starts to run out the military in every major country will grab as much of the stuff as they can and start stockpiling supplies. Not even doctors and hospitals will take precedence.

4. Heating (and cooling)

Fossil fuels help keep us warm when it is cold and help keep us cool when the weather is too hot. But the amount of energy used in heating and cooling our homes, shops and offices is vast.

When fossil fuels become too expensive to use in this way we will have huge problems (particularly with global warming making things worse). Modern homes are often built of much thinner material than pre 20th century homes and so those inside them are, despite the advantages of insulation and double glazing, still likely to suffer more than their ancestors. There isn't likely to be enough wood around for both cooking and heating.

Already, thousands of people die every year as a result of extreme heat and extreme cold. The figures are likely to rise dramatically.

The Author

Vernon Coleman is the iconoclastic author of well over a hundred books which have sold over two million copies in the UK, been translated into 23 languages and now sell in over 50 countries. His bestselling non fiction book `Bodypower' was voted one of the 100 most popular books of the 1980s/90s and was turned into two television series in the UK. The film of his novel Mrs Caldicot's Cabbage War was released early in 2003.

Vernon Coleman has written columns for numerous newspapers and magazines and has contributed over 5,000 articles, columns and reviews to hundreds of leading publications around the world. Many millions have consulted his advice lines and his website. Vernon Coleman has a medical degree, and an honorary science doctorate. He has worked for the Open University in the UK and is an honorary Professor of Holistic Medical Sciences at the Open International University based in Sri Lanka. Vernon Coleman has received lots of rather jolly awards from people he likes and respects. He is, for example, a Knight Commander of The Ecumenical Royal Medical Humanitarian Order of Saint John of Jerusalem, of the Knights of Malta and a member of the Ancient Royal Order of Physicians dedicated to His Majesty King Buddhadasa. In 2000 he was awarded the Yellow Emperor's Certificate of Excellence as Physician of the Millennium by the Medical Alternativa Institute. He worked as a GP for ten years and is still a registered medical practitioner. He has organised numerous campaigns both for people and for animals. He is a member of the MCC and can ride a bicycle and swim, though not at the same time. He loves cats, cricket (before they started painting slogans on the grass), cycling, cafes and, most of all, the Welsh Princess. He hates cruelty.

Vernon Coleman is balding rapidly and is widely disliked by members of the Establishment. He doesn't give a toss about either of these facts. Many attempts have been made to ban his books (many national publications ban all mention of them and UKIP, the political party, has, bizarrely banned his books attacking the European Union) but he insists he will keep writing them even if he has to write them out in longhand and sell them on street corners (though he hopes it doesn't come to this because he still has a doctor's handwriting). He is married to Donna Antoinette, the totally adorable Welsh Princess, and is very pleased about this.

If you found this book informative, the author would be grateful if you would post a suitable review in the usual places. Thank you.

For a list of books by Vernon Coleman please see his biography on amazon. Or visit www.vernoncoleman.com

Printed in Great Britain
by Amazon